职业教育赛教一体化课程改革系列规划教材

云计算技术与应用

YUNJISUAN JISHU YU YINGYONG

王世刚 韩明 主编
夏晶 库波 杨旭 冉柏权 副主编

中国铁道出版社有限公司
CHINA RAILWAY PUBLISHING HOUSE CO., LTD.

内 容 简 介

本书从学习 OpenStack 所需要掌握的技能出发，大量采用实际案例的方式进行内容讲解，希望读者能按照书中步骤完成 OpenStack 环境的搭建。全书内容包括 CentOS 操作系统、MySQL 基础、OpenStack 基础环境搭建、OpenStack 服务、日常运维及综合案例等内容。

本书适合作为高职高专学生的教材，也可作为云计算运维相关技术人员的参考用书。

图书在版编目（CIP）数据

云计算技术与应用/王世刚, 韩明主编. —北京：中国铁道出版社有限公司，2019.8（2020.8重印）
职业教育赛教一体化课程改革系列规划教材
ISBN 978-7-113-25810-8

Ⅰ.①云… Ⅱ.①王… ②韩… Ⅲ.①云计算-职业教育-教材 Ⅳ.①TP393.027

中国版本图书馆CIP数据核字(2019)第166015号

书　　名：	云计算技术与应用
作　　者：	王世刚　韩明
策　　划：	徐海英　　　　　　　　　　　编辑部电话：010-63551006
责任编辑：	王春霞　卢　笛
封面制作：	刘　颖
责任校对：	张玉华
责任印制：	樊启鹏

出版发行：中国铁道出版社有限公司（100054，北京市西城区右安门西街8号）
网　　址：http://www.tdpress.com/51eds/

印　　刷：北京柏力行彩印有限公司

版　　次：2019年8月第1版　2020年8月第2次印刷
开　　本：787 mm×1 092 mm　1/16　印张：12　字数：290千
书　　号：ISBN 978-7-113-25810-8
定　　价：39.00元

版权所有　侵权必究

凡购买铁道版图书，如有印制质量问题，请与本社教材图书营销部联系调换。电话：（010）63550836
打击盗版举报电话：（010）51873659

前 言

为认真贯彻落实教育部实施新时代中国特色高水平高职学校和专业群建设，扎实、持续地推进职校改革，强化内涵建设和高质量发展，落实双高计划，抓好2019年职业院校信息技术人才培养方案实施及配套建设，在湖北信息技术职业教育集团的大力支持下，武汉唯众智创科技有限公司统一规划并启动了"职业教育赛教一体化课程改革系列规划教材"（《云计算技术与应用》《大数据技术与应用Ⅰ》《网络综合布线》《物联网.NET开发》《物联网嵌入式开发》《物联网移动应用开发》），本书是"教育教学一线专家、教育企业一线工程师"等专业团队的匠心之作，是全体编委精益求精，在日复一日年复一年的工作中，不断探索和超越的教学结晶。本书教学设计遵循教学规律，涉及内容是真实项目的拆分与提炼。

本书内容围绕实现云计算技术与应用系统为中心，并适当扩展当前云计算技术与应用技术必备基本技能，以技能操作培养为中心，按照理论知识够用为度的原则来编写。

本书由武汉城市职业学院的王世刚、韩明任主编，黄冈职业技术学院的夏晶、武汉软件工程职业学院的库波、湖北生态工程职业技术学院的杨旭、武汉唯众智创科技有限公司的冉柏权任副主编。具体分工如下：韩明编写单元1和单元2；王世刚编写单元5和单元6；夏晶编写单元7；库波编写单元3；杨旭编写单元4；冉柏权编写单元8。全书由王世刚统稿。

由于编者水平有限，加之编写时间仓促，书中难免存在疏漏或不妥之处，敬请广大读者批评指正。

编 者

2019年6月

目 录

单元1　云计算的概念及发展历程 ……… 1
1.1　云计算的产生及演化 ………………… 1
 1.1.1　云计算的产生 …………………… 1
 1.1.2　云计算的概念 …………………… 2
 1.1.3　云计算的发展 …………………… 3
1.2　云计算的特点 ………………………… 4
1.3　云计算的应用 ………………………… 6
 1.3.1　云物联 …………………………… 6
 1.3.2　云安全 …………………………… 6
 1.3.3　云存储 …………………………… 7
1.4　私有云、公有云及混合云 …………… 10
 1.4.1　私有云 …………………………… 10
 1.4.2　公有云 …………………………… 11
 1.4.3　混合云 …………………………… 11
1.5　云计算的服务形式 …………………… 12
 1.5.1　基础设施即服务 ………………… 12
 1.5.2　平台即服务 ……………………… 13
 1.5.3　软件即服务 ……………………… 14
1.6　了解OpenStack ……………………… 14
 1.6.1　OpensStack的诞生及发展 ……… 14
 1.6.2　OpenStack核心项目 …………… 16
 1.6.3　OpenStack典型环境架构 ……… 17

单元2　CentOS基本环境配置 ……… 21
2.1　安装CentOS 7操作系统 …………… 21
2.2　Linux基本技能 ……………………… 29
 2.2.1　常用命令 ………………………… 29
 2.2.2　Vim编辑器的使用 ……………… 32
 2.2.3　镜像的挂载方式 ………………… 33
 2.2.4　Yum源的配置及软件包的安装 … 34
2.3　网络配置 ……………………………… 35
 2.3.1　网卡文件的配置 ………………… 35
 2.3.2　解决常见网络故障 ……………… 37

单元3　MySQL数据库中数据的基本操作 ……………………… 39
3.1　在CentOS中MySQL数据库安装及操作 … 39
 3.1.1　MySQL数据库的安装 ………… 39
 3.1.2　MySQL数据库的备份与还原 … 41
 3.1.3　MySQL数据库用户管理 ……… 41
 3.1.4　MySQL数据库的权限管理 …… 42
3.2　在MySQL数据库中插入数据 ……… 42
 3.2.1　为表中所有字段添加数据 ……… 42
 3.2.2　为表中指定字段添加数据 ……… 43
 3.2.3　同时添加多条记录 ……………… 44
3.3　在MySQL数据库中更新数据 ……… 44
3.4　在MySQL数据库中查询数据 ……… 45
 3.4.1　SELECT语法 …………………… 45
 3.4.2　简单查询 ………………………… 45
 3.4.3　条件查询 ………………………… 46
 3.4.4　结果排序 ………………………… 47
3.5　在MySQL数据库中删除数据 ……… 47

单元4 使用Python对OpenStack进行二次开发 ········· 49

- 4.1 需求分析 ········· 49
  - 4.1.1 软件项目开发流程 ········· 49
  - 4.1.2 项目需求分析 ········· 50
  - 4.1.3 对OpenStack进行二次开发需求分析 ········· 51
- 4.2 数据库设计 ········· 53
- 4.3 技能训练 ········· 54
  - 4.3.1 使用PowerDesigner进行数据库设计 ········· 54
  - 4.3.2 使用Axure进行项目原型设计 ········· 58

单元5 OpenStack基础配置 ········· 63

- 5.1 OpenStack环境准备工作 ········· 63
  - 5.1.1 OpenStack实验部署架构 ········· 63
  - 5.1.2 OpenStack实验环境硬件需求 ········· 63
  - 5.1.3 修改结点名称 ········· 65
  - 5.1.4 安全设置 ········· 65
  - 5.1.5 配置Yum源 ········· 66
- 5.2 主机网络设置 ········· 66
- 5.3 网络时间协议(NTP)设置 ········· 68
- 5.4 OpenStack包的安装 ········· 69
- 5.5 安装及设置SQL数据库 ········· 69
- 5.6 消息服务器设置 ········· 70
- 5.7 安装及设置Memcached ········· 70
- 5.8 配置Etcd ········· 71

单元6 安装OpenStack服务 ········· 73

- 6.1 身份认证服务Keystone的安装配置 ········· 73
  - 6.1.1 Keystone目录结构 ········· 74
  - 6.1.2 安装和配置组件 ········· 75
  - 6.1.3 创建Keystone的一般实例 ········· 80
  - 6.1.4 验证Keystone是否安装成功 ········· 82
  - 6.1.5 创建OpenStack客户端环境脚本 ········· 82
- 6.2 镜像服务Glance的安装配置 ········· 84
  - 6.2.1 Glance目录结构 ········· 85
  - 6.2.2 在控制端安装镜像服务Glance ········· 85
- 6.3 计算服务Nova的安装配置 ········· 94
  - 6.3.1 Nova目录结构 ········· 96
  - 6.3.2 安装和配置控制结点的计算服务 ········· 97
  - 6.3.3 验证控制结点与计算结点的计算服务 ········· 117
- 6.4 网络部署服务Neutron的安装配置 ········· 118
  - 6.4.1 Neutron目录结构 ········· 118
  - 6.4.2 安装和配置控制结点网络服务 ········· 119
  - 6.4.3 安装和配置计算结点网络服务 ········· 130
  - 6.4.4 验证网络服务 ········· 134
- 6.5 Dashboard的安装配置 ········· 136
  - 6.5.1 Dashboard 安装和配置组件 ········· 136
  - 6.5.2 验证Dashboard ········· 138
- 6.6 块存储服务Cinder的安装配置 ········· 138
  - 6.6.1 Cinder目录结构 ········· 140
  - 6.6.2 安装和配置控制结点 ········· 141
  - 6.6.3 安装和配置存储结点 ········· 144
  - 6.6.4 验证Cinder操作 ········· 146

单元7 OpenStack日常运维 ········· 147

- 7.1 控制结点的维护与排错 ········· 147
- 7.2 计算结点的维护与排错 ········· 148
- 7.3 网络诊断 ········· 148
  - 7.3.1 检查网卡状态 ········· 148
  - 7.3.2 虚拟机网络 ········· 148
  - 7.3.3 检测网络 ········· 149
- 7.4 诊断DHCP和DNS故障 ········· 149

7.4.1 日志与监控：故障的定位·········150
7.4.2 错误日志·········151
7.5 监控指标·········153
7.6 备份与恢复·········157
　7.6.1 备份的分类·········157
　7.6.2 MySQL数据库备份与恢复·········158
7.7 OpenStack常用故障处理方法·········159

单元8　综合案例·········163

8.1 OpenStack数据库详解·········163
　8.1.1 项目相关数据库·········163
　8.1.2 OpenStack中主要的数据库表·········165
8.2 OpenStack API理解·········165
　8.2.1 使用OpenStack服务的方式·········165
　8.2.2 OpenStack中的常规API类型·········166
　8.2.3 OpenStack API使用规范·········166
8.3 获取镜像列表·········168
8.4 镜像上传与编辑·········171
　8.4.1 创建镜像（镜像上传）·········171
　8.4.2 删除镜像·········172
　8.4.3 镜像修改·········173
8.5 获取云主机列表·········173
8.6 云主机相关操作·········175
　8.6.1 创建云主机·········175
　8.6.2 创建云主机类型·········175
　8.6.3 修改云主机类型·········176
　8.6.4 删除云主机类型·········176
　8.6.5 云主机运行管理·········177

单元 1

云计算的概念及发展历程

学习目标

- 了解云计算的产生；
- 了解云计算的概念；
- 了解云计算的发展历史；
- 了解云计算的特点；
- 了解云计算的应用；
- 了解公有云、私有云、混合云的区别；
- 了解 IaaS、PaaS、SaaS。

1.1 云计算的产生及演化

1.1.1 云计算的产生

传统模式下，企业建立一套 IT 系统不仅仅需要购买硬件等基础设施，还需要购买软件的许可证，需要专门的人员维护。当企业的规模扩大时还要继续升级各种软硬件设施以满足需要。对于企业来说，计算机硬件和软件本身并非他们真正需要的，硬件和软件仅仅是完成工作、提供效率的工具而已。对个人来说，人们想正常使用计算机需要安装许多软件，而许多软件是收费的，对不经常使用该软件的用户来说购买是非常不划算的。可不可以有这样的服务，能够提供人们需要的所有软件以租用？这样只需要在用时付少量"租金"即可，"租用"这些软件服务为人们节省许多购买软硬件的资金。

人们每天都要用电，但不是每家自备发电机，它由电厂集中提供；人们每天都要用自来水，但不是每家都有井，它由自来水厂集中提供。这种模式极大地节约了资源，方便人们的生活。面对计算机给人们带来的困扰，可不可以像用水和用电一样使用计算机资源？这些想法最终促使云计算的产生。

云计算的最终目标是将计算、服务和应用作为一种公共设施提供给公众，使人们能够像使用水、电、煤气和电话那样使用计算机资源。

云计算模式即为电厂集中供电模式。在云计算模式下，用户的计算机会变得十分简单，或许不大的内存、不需要硬盘和各种应用软件，就可以满足大众的需求。因为用户的计算机除了通过浏览器给"云"发送指令和接收数据外基本上什么都不用做便可以使用云服务提供商的计算资源、存储空间和各种应用软件。这就像连接"显示器"和"主机"的电线无限长，从而可以把"显示器"放在使用者的面前，而"主机"放在远到甚至计算机使用者本人也不知道的地方。

云计算把连接"显示器"和"主机"的电线变成网络，把"主机"变成云服务提供商的服务器集群。

在云计算环境下，用户的使用观念也会发生彻底变化：从"购买产品"向"购买服务"转变，因为他们直接面对的将不再是复杂的硬件和软件，而是最终的服务。用户不需要拥有看得见、摸得着的硬件设施，也不需要为机房支付设备供电、空调制冷、专人维护等费用，并且不需要等待漫长的供货周期、项目实施等冗长的时间，只要把钱汇给云计算服务提供商，就会马上得到需要的服务。

1.1.2 云计算的概念

云计算（Cloud Computing）是一种基于互联网的计算新方式，通过互联网上异构、自治的服务为个人和企业用户提供按需即取的计算。由于资源存储在互联网上，互联网常以一个云状图案来表示，因此可以形象地类比为云，"云"同时也是对底层基础设施的一种抽象概念。

云计算是分布式处理（Distributed Computing）、并行处理（Parallel Computing）和网格计算（Grid Computing）的发展，或者说是这些计算机科学概念的商业实现。许多跨国信息技术行业的公司如IBM、Yahoo 和 Google 等正在使用云计算的概念销售自己的产品和服务。

云计算的资源是动态易扩展而且虚拟化的，通过互联网提供。终端用户不需要了解"云"中基础设施的细节，不必具备相应的专业知识，也无须直接进行控制，只关注自己真正需要什么样的资源，以及如果通过网络得到相应的服务。

云计算可以认为包括以下几个层次的服务：基础设施及服务（IaaS）、平台及服务（PaaS）和软件及服务（SaaS）。云计算服务通常提供通用的通过浏览器访问的在线商业应用，软件和数据可存储在数据中心。

互联网上的云计算服务特征与自然界的云、水循环具有一定的相似性，因此，云是一个相当贴切的比喻。通常云计算服务应具备以下几条特征：基于虚拟化技术快速部署资源或获得服务；实现动态的、可伸缩的扩展；按需提供资源、按使用量付费；通过互联网提供、面向海量信息处理；用户可以方便地参与。云计算服务框架如图 1-1 所示。

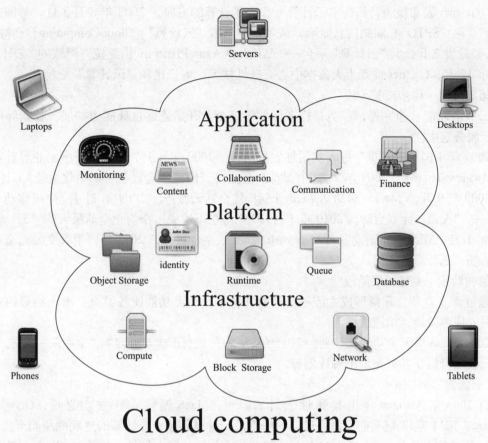

图 1-1 云计算服务框架

1.1.3 云计算的发展

"云计算"的变更和发展基本可以分为四个阶段：

第一个阶段：理论完善阶段

SaaS/IaaS 云服务出现，并被市场接受。

1959 年 6 月，ChristopherStrachey 发表虚拟化论文，虚拟化是今天"云计算"基础架构的基石。1984 年，Sun 公司的联合创始人 John Gage 提出了"网络就是计算机"的名言，用于描述分布式计算技术带来的新世界，今天的"云计算"正在将这一理念变成现实。1997 年，南加州大学教授 Ramnath K.Chellappa 提出"云计算"的第一个学术定义，认为计算的边界可以不是技术局限，而是经济合理性。1999 年，MarcAndreessen 创建 LoudCloud，是第一个商业化的 IaaS 平台。1999 年 3 月 Salesforce 成立，成为最早初现的云服务，即 SaaS 服务。2005 年，Amazon 宣布 AmazonWebServices "云计算"平台。

第二阶段：发展准备阶段

云服务的三种形式全部出现，IT 企业、电信运营商、互联网企业等纷纷推出云服务，云服务形成。

2007 年，Salesforce 发布 Force.com 即 PaaS 服务。2007 年 11 月，IBM 首次发布"云计算"商业解决方案，推出"蓝云"（BlueCloud）计划。2008 年 4 月，GoogleAppEngine 发布。2008 年

年中，Gartner 发布报告，认为"云计算"代表了计算的方向。2008 年 8 月 3 日，美国专利商标局（以下简称"SPTO"）网站信息显示，戴尔正在申请"云计算"（CloudComputing）商标。2008 年 10 月，微软发布其公共"云计算"平台——WindowsAzurePlatform，由此拉开微软的"云计算"大幕。2008 年 12 月，Gartner 披露十大数据中心突破性技术，虚拟化和"云计算"上榜。

第三阶段：稳步成长阶段

云服务功能日趋完善，种类日趋多样；传统企业开始通过自身能力扩展、收购等模式、纷纷投入云服务之中。

2009 年中国"云计算"进入实质性发展阶段。2009 年 4 月，VMware 推出业界首款云操作系统 VMwarevSphere4。2009 年 7 月，中国首个企业"云计算"平台诞生——中化企业"云计算"平台。

2009 年 9 月，VMware 启动 vCloud 计划构建全新云服务。2009 年 11 月，中国移动"云计算"平台——"大云"计划启动。2010 年 1 月，IBM 与松下达成迄今为止全球最大的"云计算"交易。2010 年 1 月，Microsoft 正式发布 MicrosoftAzure 云平台服务。2013 年，甲骨文公司全面展示了甲骨文最新"云计算"产品。

第四阶段：高速发展阶段

通过深度竞争，逐渐形成主流平台产品和标准；产品功能比较健全、市场格局相对稳定；云服务进入成熟阶段，增速放缓。

2014 年，阿里云启动云合计划。2015 年，华为在北京正式对外宣布"企业云"战略。2016 年，腾讯云战略升级，并宣布云出海计划等。

"云计算"的由来

自 2005 年 Amazon 推出其弹性云计算，并在 IaaS 领域大获成功之后，Google、微软、VMware、IBM 等 IT 巨头们大举跟进，纷纷涉足云计算领域，云计算的热潮汹涌而至。

从最早的 Eucalyptus、OpenNebula、CloudStack 等，到人们熟知的 OpenStack，百花齐放，百家争鸣，一时云计算领域热闹非凡。当然，随着众多大型厂商的支持、社区的壮大以及生态体系的不断完善，OpenStack 如今大放异彩，开始逐渐占据主导地位。不是说 OpenStack 现在完美无缺，只是在当下的开源 IaaS 领域，OpenStack 已经毫无争议地成为最受关注的云计算技术，逐渐成为开源 IaaS 领域的代名词。

在 PaaS 层，基于标准的开源 PaaS 有：Cloudify（GigaSpaces）；厂商驱动的开源 PaaS 有：Cloud Foundry（原先由 VMware 开发）、OpenShift（红帽）、Stackato（ActiveState）等技术，都已成为国内云计算 PaaS 层开源的主流。

除此之外，在云计算的关键技术领域，开源的力量也无处不在，Kvm 虚拟化技术、Ceph 存储技术、Container 技术在开源社区独树一帜，社区空前活跃，影响力也不容小觑，生态体系建设俨然成形。

1.2 云计算的特点

云计算的特点有：

1. 超大规模

"云"具有相当的规模，Google 云计算已经拥有 100 多万台服务器，Amazon、IBM、微软、Yahoo 等的"云"均拥有几十万台服务器。企业私有云一般拥有数百上千台服务器。"云"能赋予用户前所未有的计算能力。

2. 虚拟化

云计算支持用户在任意位置、使用各种终端获取应用服务。所请求的资源来自"云"，而不是固定的、有形的实体。应用在"云"中某处运行，但实际上用户无须了解、也不用担心应用运行的具体位置。只需要一台笔记本计算机或者一部手机，就可以通过网络服务来实现人们需要的一切，甚至包括超级计算这样的任务。

3. 高可靠性

"云"使用了数据多副本容错、计算结点同构可互换等措施来保障服务的高可靠性，使用云计算比使用本地计算机可靠。

4. 通用性

云计算不针对特定的应用，在"云"的支撑下可以构造出千变万化的应用，同一个"云"可以同时支撑不同的应用运行。

5. 高可扩展性

"云"的规模可以动态伸缩，满足应用和用户规模增长的需要。

6. 按需服务

"云"是一个庞大的资源池，用户按需购买；云可以像自来水、电、煤气那样计费。

7. 极其廉价

由于"云"的特殊容错措施可以采用极其廉价的结点来构成云，"云"的自动化集中式管理使大量企业无须负担日益高昂的数据中心管理成本，"云"的通用性使资源的利用率较之传统系统大幅提升，因此用户可以充分享受"云"的低成本优势，经常只要花费几百美元、几天时间就能完成以前需要数万美元、数月时间才能完成的任务。

云计算可以彻底改变人们未来的生活，但同时也要重视环境问题，这样才能真正为人类进步做贡献，而不是简单的技术提升。

8. 潜在的危险性

云计算服务除了提供计算服务外，必然还提供存储服务。但是云计算服务当前垄断在私人机构（企业）手中，而他们仅仅能够提供商业信用。政府机构、商业机构（特别像银行这样持有敏感数据的商业机构）对于选择云计算服务应保持足够的警惕。一旦商业用户大规模使用私人机构提供的云计算服务，无论其技术优势有多强，都不可避免地让这些私人机构以"数据（信息）"的重要性挟制整个社会。对于信息社会而言，"信息"是至关重要的。另一方面，云计算中的数据对于数据所有者以外的其他云计算用户是保密的，但是对于提供云计算的商业机构而言确实毫无秘密可言。所有这些潜在的危险，是商业机构和政府机构选择云计算服务，特别是国外机构提供的云计算服务时，不得不考虑的一个重要的前提。

1.3 云计算的应用

1.3.1 云物联

什么是云物联？"云"怎么和"物联"关联在一起？什么关系？什么理念？什么技术？什么工具？什么意思？有什么用？其实，"云"指云计算，"物联"指物联网。要想很好理解云物联，必须对于云计算和物联网及其二者的关系有一个完整且完善的把握。

什么是物联网？其实对于物联网，人们恐怕接触的 RFID、传感技术等都不会落下物联网的影子。从技术上来讲，物联网又有什么特点呢？从目前公众比较认可的理解来看，物联网是指在现实世界中，那些具有一定的感知能力、识别能力或者具有其他智能化特点的传感设备。除了刚才提到的传感器、RFID 以外，像二维码、小范围的无线传输技术，或者一些移动通信的模块等，都可以算在范围内。总的来说，物联网技术基本就是通过网络设施实现信息的传输和管理的技术，从而让人们可以打破空间距离的限制，更好的实现隔空交流或者设备的控制。严格意义上来说，物联网应该被划为工程学的分类当中。原因在于它强调的是在互联网的基础上，物与物之间建立的一种连接关系。如果从作用的角度来分析，物联网技术带给事物的是一种技术上的革新，而它的载体还是原有的事物，并不是一些人理解的取代，如果真要取代，那么物联网赋予物品先进的意义恐怕也会随之消失。

云计算与物联网之间有何关系？由于很多时候云计算和物联网经常在同一个场景中出现，所以很多人认为二者之间有着密切的联系。其实从技术的角度来说，物联网和云计算还着实没有什么太多的联系。如果从工作架构的角度分析，物联网则可以认为是承载云计算技术的一个平台。

借助云计算技术的支持，物联网则可以更好地提升数据的存储以及处理能力。从而使自身的技术得到进一步完善。如果失去云计算的支持，物联网的工作性能无疑会大打折扣，而在和其他传统的技术相比，它的意义也会大大降低。所以说物联网对云计算有着很强的依赖性。

但是对于云计算技术来说，物联网只是其众多的应用载体之一。尽管物联网对云计算有着很强的依赖，但作为诸多载体的一种，云计算对于物联网可真没有"特殊的感情"。甚至就像柜台与货物的关系一样，货物在哪个台子上都会使这个柜台受到更多的关注；如果柜台失去了货物，那么，无论是作用还是关注度都会随之降低。因此，云计算对于物联网只是普通的"合作关系"，并没有什么太特殊的联系。

不过，从目前的发展前景来看，物联网和云计算之间的联系将会向着越来越紧密的方向发展。在云计算技术的支持下，被赋予更强工作能力的物联网，在我国的使用率呈现逐年递增的趋势，所涉及的领域也越来越广泛。我国的 RFID 技术已经成为世界排名第三的应用市场，当前被各个领域广泛应用的二维码识别技术，其实也是物联网 RFID 技术中的一种。所以说，云计算承载物联网在我国有着更加广阔的发展空间。

1.3.2 云安全

"云安全（Cloud Security）"计划是网络时代信息安全的最新体现，它融合了并行处理、网格计算、未知病毒行为判断等新兴技术和概念，通过网状的大量客户端对网络中软件行为的异常监测，获取互联网中木马、恶意程序的最新信息，传送到 Server 端进行自动分析和处理，再把病毒和木马的解决方案分发到每一个客户端。

"云安全"是继"云计算""云存储"之后出现的"云"技术的重要应用,已经在反病毒软件中取得了广泛应用,发挥了良好的效果。在病毒与反病毒软件的技术竞争当中为反病毒软件夺得了先机。

未来杀毒软件将无法有效地处理日益增多的恶意程序。来自互联网的主要威胁正在由计算机病毒转向恶意程序及木马,在这样的情况下,采用的特征库判别法显然已经过时。云安全技术应用后,识别和查杀病毒不再仅仅依靠本地硬盘中的病毒库,而是依靠庞大的网络服务,实时进行采集、分析及处理。整个互联网就是一个巨大的"杀毒软件",参与者越多,每个参与者就越安全,整个互联网就会更安全。

云安全技术是 P2P 技术、网格技术、云计算技术等分布式计算技术混合发展、自然演化的结果。

要想建立"云安全"系统,并使之正常运行,需要解决云安全系统的四大难点问题:

第一,需要海量的客户端(云安全探针)。只有拥有海量的客户端,才能对互联网上出现的恶意程序、危险网站有最灵敏的感知能力。一般而言安全厂商的产品使用率越高,反应应当越快,最终应当能够实现无论哪个网民中毒、访问挂马网页,都能在第一时间做出反应。

第二,需要专业的反病毒技术和经验。发现的恶意程序被探测到,应当在尽量短的时间内被分析,这需要安全厂商具有过硬的技术,否则容易造成样本的堆积,使云安全快速探测的结果大打折扣。

第三,需要大量的资金和技术投入。"云安全"系统在服务器、带宽等硬件方面需要极大的投入,同时要求安全厂商应当具有相应的顶尖技术团队、持续的研究经费。

第四,可以是开放的系统,允许合作伙伴加入。"云安全"可以是个开放性的系统,其"探针"应当与其他软件相兼容,即使用户使用不同的杀毒软件,也可以享受"云安全"系统带来的成果。

1.3.3 云存储

1. 云存储的概念

云存储(Cloud Storage)这个概念一经提出,就得到了众多厂商的支持和关注。Amazon 推出的 Elastic Compute Cloud(EC2:弹性计算云)云存储产品,旨在为用户提供互联网服务形式的同时提供更强的存储和计算功能。内容分发网络服务提供商 CDNetworks 和业界著名的云存储平台服务商 Nirvanix 发布了一项新的合作,并宣布结成战略伙伴关系,以提供业界目前唯一的云存储和内容传送服务集成平台。微软推出提供网络移动硬盘服务的 Windows Live SkyDrive Beta 测试版。EMC 宣布加入道里可信基础架构项目,致力于云计算环境下关于信任和可靠度保证的全球研究协作,IBM 也将云计算标准作为全球备份中心的 3 亿美元扩展方案的一部分。

云存储变得越来越热,大家众说纷"云",而且各有各的说法,各有各观点,那么到底什么是云存储?

云存储是在云计算(Cloud Computing)概念上延伸和发展出来的一个新的概念。云计算是分布式处理(Distributed Computing)、并行处理(Parallel Computing)和网格计算(Grid Computing)的发展,是透过网络将庞大的计算处理程序自动分拆成无数个较小的子程序,再交由多台服务器所组成的庞大系统经计算分析之后将处理结果回传给用户。通过云计算技术,网络服务提供者可以在数秒之内,处理数以千万计甚至亿计的信息,达到和"超级计算机"同样强大的网络服务。

云存储的概念与云计算类似，它是指通过集群应用、网格技术或分布式文件系统等功能，将网络中大量各种不同类型的存储设备通过应用软件集合起来协同工作，共同对外提供数据存储和业务访问功能的一个系统。

如果这样解释还是难以理解，那么可以借用广域网和互联网的结构来解释云存储。

2. 云状的网络结构

相信大家对局域网、广域网和互联网都已经非常了解了。在常见的局域网系统中，为了能更好地使用局域网，一般来讲，使用者需要非常清楚地知道网络中每一个软硬件的型号和配置，如采用什么型号的交换机，有多少个端口，采用了什么路由器和防火墙，分别是如何设置的等。系统中有多少台服务器，分别安装了什么操作系统和软件。但当使用广域网和互联网时，只需要知道是什么样的接入网和用户名、密码就可以连接到广域网和互联网，并不需要知道广域网和互联网中到底有多少台交换机、路由器、防火墙和服务器，不需要知道数据是通过什么样的路由到达用户的计算机，也不需要知道网络中的服务器分别安装了什么软件，更不需要知道网络中各设备之间采用了什么样的连接线缆和端口。

广域网和互联网对于具体的使用者是完全透明的，经常用一个云状的图形来表示广域网和互联网。

虽然云状的图形中包含许许多多的交换机、路由器、防火墙和服务器，但对具体的广域网、互联网用户来讲，这些都是不需要知道的。这个云状图形代表的是广域网和互联网带给大家的互联互通的网络服务，无论在什么地方，都可以通过一个网络接入线缆和一个用户名、密码，即可接入广域网和互联网，享受网络带来的服务。

参考云状的网络结构，创建一个新型的云状结构的存储系统，这个存储系统由多台存储设备组成，通过集群功能、分布式文件系统或类似网格计算等功能联合起来协同工作，并通过一定的应用软件或应用接口，对用户提供一定类型的存储服务和访问服务。

当人们使用某一个独立的存储设备时，必须非常清楚这个存储设备是什么型号、什么接口和传输协议，必须清楚地知道存储系统中有多少块磁盘，分别是什么型号、多大容量，必须清楚存储设备和服务器之间采用什么样的连接线缆。为了保证数据安全和业务的连续性，还需要建立相应的数据备份系统和容灾系统。除此之外，对存储设备进行定期的状态监控、维护、软硬件更新和升级也是必需的。

如果采用云存储，那么上面所提到的一切对使用者来讲都不需要了。云状存储系统中的所有设备对使用者来讲都是完全透明的，任何地方的任何一个经过授权的使用者都可以通过一根接入线缆与云存储连接，对云存储进行数据访问。

云存储不是存储，而是服务。就如同云状的广域网和互联网一样，云存储对使用者来讲，不是指某一个具体的设备，而是指一个由许许多多存储设备和服务器所构成的集合体。使用者使用云存储，并不是使用某一个存储设备，而是使用整个云存储系统带来的一种数据访问服务。严格来讲，云存储不是存储，而是一种服务。

云存储的核心是应用软件与存储设备相结合，通过应用软件来实现存储设备向存储服务的转变。

3. 云存储的优势

不得不说云存储是一种极具诱惑的存储技术，功能强大，灵活多变。

1) 设备层面

云存储的存储设备数量庞大，分布区域各异，多台设备之间进行协同合作，许多设备可以同时为某一个人提供同一种服务，并且云存储都是平台服务，云存储的供应商会根据用户需求开发多种多样的平台，如 IPTV 应用平台、视频监控应用平台、数据备份应用平台等。只要有标准的公用应用接口，任何一个被授权的用户都可以通过一个简单的网址登录云存储系统，享受云存储服务。

2) 功能层面

云存储的容量分配不受物理硬盘的控制，可以按照客户的需求及时扩容，设备故障和设备升级都不会影响用户的正常访问。云存储技术针对数据重要性采取不同的复制策略，并且复制的文件存放在不同的服务器上，因此当硬件损坏时，不管是硬盘还是服务器，服务始终不会终止，而且正因为采用索引的架构，系统会自动将读/写指令引导到其他存储结点，读/写效能完全不受影响，管理人员只要更换硬件即可，数据也不会丢失。换上新的硬盘服务器后，系统会自动将文件复制回来，永远保持多备份的文件，从而避免数据丢失。在扩容时，只要安装好存储结点，接上网络，新增加的容量便会自动合并到存储中，其数据会自动迁移到新存储的结点，不需要做多余的设定，大大降低了维护人员的工作量。

3) 开支层面

传统存储模式一旦完成资金的一次性投入，系统无法在后续使用中动态调整。随着设备的更新换代，落后的硬件平台难以处置；随着业务需求不断变化，软件需要不断地更新升级甚至重构来与之相适应，导致维护成本高昂，很容易发展到不可控的程度。但使用云存储服务，可以免去企业在设备购买和技术人员聘用上的庞大开支，至于维护工作以及系统的更新升级都由云存储服务提供商完成，而且公用云的租用费用和私有云的建设费用会随着云存储供应商竞争的日趋激烈而不断减低。云存储是未来的存储应用趋势。

4. 云存储的劣势

作为新生事物，云存储的优势有目共睹，但是不可否认，云存储也有劣势，就目前情况来看，业界内成功的云存储服务是少之又少。

1) 安全问题

云存储的好处在于只要有标准的公用应用接口，任何一个被授权的用户都可以访问云存储系统，查看相关数据，但这种便利性不能不说也是云存储的致命伤。因为每一种类的设备都有每一种设备的可攻击点，倘若用户借助手机端口访问云存储系统，而恰巧该用户在使用过程中数据被拦截了，那结果显而易见，数据极有可能被泄露。虽然目前许多云存储都采用加密或者其他的安全技术，但是，这些安全技术并不能把云存储装扮成铜墙铁壁，除非进行二次校验或者二次加密。但进行二次验证和加密无疑又加大了供应商的开发难度和用户访问相关数据的烦琐性。

2) 访问速度问题

访问速度慢这是当前云存储短时间内无法突破的一个瓶颈，也是许多用户诟病的地方。截至目前，云存储还不能处理交易相对频繁的文件，需要快速的网络连接的数据库也不是云存储的存储对象，Tier1、Tier2 或以块为基础的数据存储亦是超出了云存储的存储能力。只有一些庞大的档案资料和非结构化数据适合云存储消化，如银行的开户信息、过去一段时间的账户交易信息，医疗机构的病患资料和病史资料等。目前访问速度慢的主要根源在于云存储提供商所提供设备的性能和带宽因素。相信这两点在不久的将来都会得到解决。

3）数据所有权问题

云存储的主要功能是借助大型的存储设备将相同的数据分别存储在不同的地域，形成数据备份，帮用户解决容灾难题。虽然用户通过与供应商签订服务水平协议免去了数据丢失甚至数据遭受破坏的后顾之忧，但从另一方面来说，用户的知情权却少得可怜，他们只知道自身的数据存储在云存储当中，但是至于存储在什么地方，有没有在没有授予权限的情况下被别人访问了就不得而知了，知识产权得不到相应的保护，数据所有权得不到相应的保障。

1.4 私有云、公有云及混合云

1.4.1 私有云

私有云（Private Clouds）是为一个客户单独使用而构建的，因而提供对数据、安全性和服务质量的最有效控制。私有云的基础是首先你要拥有基础设施并可以控制在此设施上部署应用程序的方式，私有云可部署在企业数据中心的防火墙内，也可以将它们部署在一个安全的主机托管场所。私有云可以搭建在公司的局域网上，与公司内部的公司监控系统、资产管理系统等相关系统进行连通，从而有利于公司内部系统的集成管理。

私有云可由公司自己的 IT 机构建，也可由云提供商进行构建。在此"托管式专用"模式中，像 Sun、IBM 这样的云计算提供商可以安装、配置和运营基础设施，以支持一个公司企业数据中心内的专用云。此模式赋予公司对于云资源使用情况的极高水平的控制能力，同时带来建立并运作该环境所需的专门知识。一个著名的技术专家曾经说过这样的一句话："对于云计算，每个人都有自己的定义。"但是大家对于云计算很多方面都有一定的共识，如三层架构（SaaS、PaaS 和 IaaS）。除了三层架构外，大家也都认为云可以被分为三种：公有云、私有云和混合云。

部署私有云的优点：

（1）数据安全。虽然每个公有云的提供商都对外宣称，其服务在各方面都是非常安全的，特别是对数据的管理。但是对企业而言，特别是大型企业而言，和业务有关的数据是其生命线，不能受到任何形式的威胁，所以短期而言，大型企业是不会将其 Mission-Critical 的应用放到公有云上运行。而私有云在这方面是非常有优势的，因为它一般都构筑在防火墙后。

（2）SLA（服务质量）。因为私有云一般在防火墙之后，而不是在某一个遥远的数据中心，所以当公司员工访问那些基于私有云的应用时，它的 SLA 应该会非常稳定，不会受到网络不稳定的影响。

（3）充分利用现有硬件资源和软件资源。大家知道每个公司，特别大公司都会有很多 legacy 的应用，而且 legacy 大多都是其核心应用。虽然公有云的技术很先进，但却对 legacy 的应用支持不好，因为很多都是用静态语言编写的，以 COBOL、C、C++ 和 Java 为主，而现有的公有云对这些语言的支持很一般。但私有云在这方面就不错，如 IBM 推出的 cloudburst，通过 cloudburst 能非常方便地构建基于 Java 的私有云。而且一些私有云的工具能够利用企业现有的硬件资源来构建云，这样将极大降低企业的花费。

（4）不影响现有 IT 管理的流程。对大型企业而言，流程是其管理的核心，如果没有完善的流程，企业将会成为一盘散沙。不仅有与业务相关的流程非常繁多，而且 IT 部门的流程也不少，如那些和 Sarbanes-Oxley 相关的流程，并且这些流程对 IT 部门非常关键。此时使用公有云很吃亏，假如使用公有云，将会对 IT 部门流程有很多冲击，如在数据管理方面和安全规定等方面。

1.4.2 公有云

公有云通常指第三方提供商为用户提供的能够使用的云，公有云一般可通过 Internet 使用，可能是免费或成本低廉的。这种云有许多实例，可在当今整个开放的公有网络中提供服务。能够以低廉的价格，提供有吸引力的服务给最终用户，创造新的业务价值，公有云作为一个支撑平台，还能够整合上游的服务（如增值业务、广告等）提供者和下游最终用户，打造新的价值链和生态系统。

公有云被认为是云计算的主要形态。目前在国内发展如火如荼，根据市场参与者类型分类，可以分为五类：

(1) 传统电信基础设施运营商，包括中国移动、中国联通和中国电信；
(2) 政府主导下的地方云计算平台，如各地的各种"××云"项目；
(3) 互联网巨头打造的公有云平台，如阿里云、华为云等；
(4) 部分 IDC 运营商，如世纪互联；
(5) 具有国外技术背景或引进国外云计算技术的国内企业，如亚马逊 AWS 等。

公有云的计算模型分为三个部分：

1. 公有云接入

个人或企业可以通过普通的互联网来获取云计算服务，公有云中的"服务接入点"负责对接入的个人或企业进行认证，判断权限和服务条件等，通过"审查"的个人和企业，就可以进入公有云平台并获取相应的服务。

2. 公有云平台

公有云平台是负责组织协调计算资源，并根据用户的需要提供各种计算服务。

3. 公有云管理

公有云管理对"公有云接入"和"公有云平台"进行管理监控，它面向的是端到端的配置、管理和监控，为用户可以获得更优质的服务提供保障。

1.4.3 混合云

混合云是私有云、公有云集中的任意混合，这种混合可以是计算的、存储的，也可以两者兼而有之。在公有云尚不完全成熟，而私有云存在运维难、部署实践长、动态扩展难的现阶段，混合云是一种较为理想的平滑过渡方式，短时间内的市场占有率将会大幅上升，并且不混合是相对的，混合是绝对的。在未来，即使不是自家的私有云和公有云混合，也需要内部的数据与服务和外部的数据与服务进行不断的调用（PaaS 级混合）。

混合云综合了数据安全性以及资源共享性双方面的考虑，个性化的方案达到省钱、安全的目的，从而获得越来越多企业的青睐。但混合云也并不是完美无缺的，以下几个问题需要格外注意。

(1) 数据冗余能力：混合云缺少数据冗余，对于数据而言，做好冗余以及容灾备份是非常必要的，若缺乏数据冗余能力，实际上数据安全性也不能得到很好的保证。

(2) 法律方面：混合云是公有云和私有云的集合，因此在法律法规上必须确保公有云和私有云的提供符合法律规范，而且必须要证明两个云之间是顺从的。

（3）SLA（服务质量）：相比私有云而言有可能会略差，在这里 SLA 指的是标准统一性。集成公有云和私有云寻求潜在的问题都会破坏服务。比如：如果一个私有云的关键业务驱动在本地保持敏感和机密数据，然后 SLA 应该体现出在公有云中使用这些服务的限制性。

（4）风险成本或学习成本较高。从安全角度而言，混合云虽然具备私有云的安全性，但是随之带来的是由于 API 带来的复杂网络配置使得传统管理员的知识经验及能力受到挑战，以及高昂的学习成本或者系统管理员能力不足带来的额外风险。

1.5 云计算的服务形式

1.5.1 基础设施即服务

云计算基础设施是内部系统和公共云之间的软件和硬件层，其融合许多不同的工具和解决方案，是成功实现云计算部署的重要系统。

随着公共云改变了数据中心及其硬件的结构，这一层次的云计算基础设施不断发展。到目前为止，IT 设备和数据中心系统采用更加谨慎的方法，一切设施都在防火墙后面。只有用户的应用和数据在企业内部和防火墙内部，其应用程序也是如此。

如今，企业的业务需要面向外部厂商，如 AWS、Azure、谷歌云或其他云计算公司。企业需要在其防火墙中创建安全的数据流，以安全地连接到公有云并防止入侵者的进入和攻击，同时保持可接受的性能水平。

1. 云计算基础设施构建模块

云计算基础设施的组件通常分为三大类：计算、网络和存储。

（1）计算：执行云系统的基本计算。这几乎是虚拟化的，因此可以移动实例。

（2）网络：通常是商用硬件运行某种软件定义网络（SDN）软件来管理云连接。

（3）存储：通常是硬盘和闪存存储的组合，旨在公有云和私有云之间来回移动数据。

存储是云基础架构与传统数据中心基础架构相分离的地方。云基础架构通常使用本地连接的存储而不是存储区域网络上的共享磁盘阵列。AWS、Azure 和 Google 等云服务提供商对 SSD 存储的收费高于硬盘存储收费。

云存储还使用为不同类型的存储方案设计的分布式文件系统，如对象、大数据或块存储。使用的存储类型取决于企业需要处理的任务，关键点在于：云存储可以根据需要扩展或缩减。

云计算基础设施是任何平台和应用程序的基础。诸如笔记本式计算机、电话或服务器之类的连接设备在这个更大的云计算系统中传输数据。

2. IaaS 的好处

IaaS 是构建云计算基础设施的基础。云计算基础设施是实体，IaaS 是商店。IaaS 使得从公有云提供商通过全球互联网租用这些云计算基础设施组件（计算、存储和网络）成为可能。

IaaS 的好处如下：

（1）削减前期成本：IaaS 消除了购买服务器硬件的前期资本支出，而这些服务器可能等待数周才能交付，需要更多时间安装和部署，最后进行配置。用户可以在 15 min 内登录公司公共云的控制面板并启动虚拟实例。

(2) 可扩展容量：如果企业需要更多容量，可以快速购买更多容量，并且如果发现不需要分配更多的容量，则可以缩减，不用支付购买新设备的前期资本支出。IaaS 遵循基于使用的消费模式，企业可以按使用容量支付费用。

(3) 折扣：IaaS 供应商还提供持续使用的折扣，或者如果企业进行大量的前期购买。节省的费用可能高达 75%。

IaaS 的下一步是平台即服务（PaaS），它建立在相同的 IaaS 平台和硬件之上。但 PaaS 已被扩展以提供更多服务，如完整的开发环境，其中包括 Web 服务器、工具、编程语言和数据库。

3. 为何使用云计算基础设施

在传统的 IT 基础设施中，一切都与服务器相关联。企业的存储数据位于特定存储阵列上，应用程序在专用物理服务器上，如果有什么事情发生，那么企业的工作就会停止。

在云计算基础设施中，因为一切都是虚拟化的，所以没有任何东西与特定的物理服务器相关联，这适用于服务和应用程序。人们是否认为当登录 Gmail 时，每次都登录到同一台物理服务器？并不是，它可能是谷歌数据中心几十个虚拟化服务器之一。

如果企业为内部基础设施部署云计算基础设施模型，这同样适用于其 AWS 实例和内部服务。通过虚拟化存储、计算和网络组件，企业可以从任何可用的服务构建，而不是大量使用。例如：企业可以在具有低利用率的硬件上的虚拟服务器上启动应用程序。或者，可以在流量较低的交换机上部署网络连接。

借助云计算基础设施，DevOps 团队可以构建他们的应用程序，以便以编程方式部署应用程序。可以告诉应用程序查找低利用率服务器或尽可能靠近数据存储部署。在传统的 IT 环境中，则无法做到这一点。

1.5.2 平台即服务

平台即服务（Platform as a Service，PaaS）是把应用服务的运行和开发环境作为一种服务提供的商业模式。在云计算的典型层级中，平台即服务层介于软件即服务与基础设施即服务之间。PaaS 是一种无须下载或安装，即可通过因特网发送操作系统和相关服务的模式。平台即服务是软件即服务（Software as a Service，SaaS）的延伸。软件即服务是将软件部署为托管服务并通过因特网提供给客户。

平台即服务提供用户能将云基础设施部署与创建至客户端，或者借此获得使用编程语言、程序库与服务。用户不需要管理与控制云基础设施，包含网络、服务器、操作系统或存储，但需要控制上层的应用程序部署与应用代管的环境。

所谓 PaaS 实际上是指将软件研发的平台作为一种服务，以 SaaS 的模式提交给用户。因此，PaaS 也是 SaaS 模式的一种应用。但是，PaaS 的出现可以加快 SaaS 的发展，尤其是加快 SaaS 应用的开发速度。PaaS 之所以能够推进 SaaS 的发展，主要在于它能够提供企业进行定制化研发的中间件平台，同时涵盖数据库和应用服务器等。PaaS 可以提高在 Web 平台上利用的资源数量。用户或者厂商基于 PaaS 平台可以快速开发自己所需要的应用和产品。同时，PaaS 平台开发的应用能更好地搭建基于 SOA 架构的企业应用。此外，PaaS 对于 SaaS 运营商来说，可以帮助他进行产品多元化和产品定制化。

1.5.3 软件即服务

软件即服务（Software as a Service，SaaS）是随着互联网技术的发展和应用软件的成熟，于 21 世纪开始兴起的一种完全创新的软件应用模式。它是云计算领域发展最成熟、应用最广泛的服务。它是一种通过互联网，为用户提供软件及应用程序的服务方式。由于基于 SaaS 的软件只有在用户需要时才被使用，SaaS 又称为"按需"软件。SaaS 模式大大降低了软件，尤其是大型软件的使用成本，并且由于软件是托管在服务提供商服务器上，减少了客户的管理维护成本，可靠性也更高。Salesforce 是 SaaS 模式的典型代表。

1.6 了解 OpeStack

1.6.1 OpenStack 的诞生及发展

1. OpenStack 发展过程

2010 年 7 月，RackSpace 和美国国家航空航天局合作，分别贡献 RackSpace 云文件平台代码和 NASA Nebula 平台代码，OpenStack 由此诞生（Austin 版）。

2011 年 2 月，OpenStack 社区发布 Bexar 版，新增 Glance 提供镜像服务。

2011 年 4 月，OpenStack 社区发布 Cactus 版，并未增加任何项目。

2011 年 9 月，OpenStack 发布第四个版本 Diablo。

2012 年 4 月，OpenStack 发布第五个版本 Essex，吸收两个新核心项目——用于用户界面操作的 Horizon 和认证的 Keystone。

2012 年 9 月，OpenStack 将 Nova 项目中的网络模块和块存储模块剥离出来，成立 Quantum 和 Cinder，并发行第六个版本 Folsom。

2013 年 4 月，OpenStack 发布第七个版本 Grizzly。

2013 年 10 月，OpenStack 发布第八个版本 Havana，首次提出集成项目的概念，并集成两个新项目——用于监控与计费的 Ceilometer 和用于编配的 Heat。

2014 年 4 月，OpenStack 发布第九个版本 Icehouse，并加入 Trove 来提供数据服务。

2014 年 10 月，第十个版本 Juno 发布。

2015 年 4 月，第十一个版本 Kilo 发布。

2015 年 10 月，第十二个版本 Liberty 发布。

2015 年 4 月，第十三个版本 Mikata 发布。

2. OpenStack 体系结构

云计算有三种落地方式：

（1）IaaS（基础架构即服务）：通过互联网提供"基础的计算资源"，包括处理能力、存储空间、网络等。

（2）PaaS（平台即服务）：把计算环境、开发环境等平台作为一种服务通过互联网提供给用户。

（3）SaaS（软件即服务）：通过互联网，为用户提供软件及应用程序的一种服务方式。

PaaS 和 SaaS 并不一定需要底层有虚拟化技术的支持，但 IaaS 一般都是建立在虚拟化技术的基础之上。OpanStack 及它一直跟随的榜样 AWS 都属于 IaaS 的范畴。

AWS 架构由五层组成，自下而上分别是 AWS 全球基础架构、基础服务、应用平台服务、管理以及用户应用程序，如图 1-2 所示。

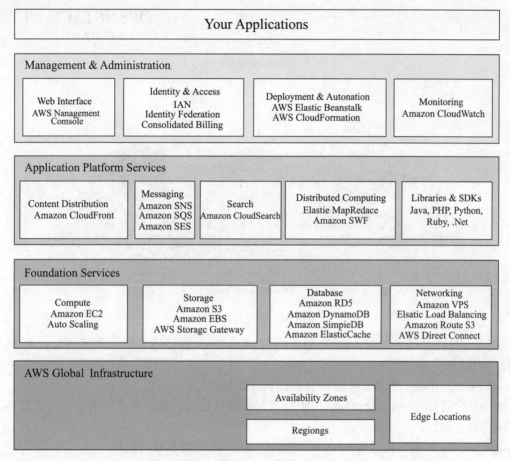

图 1-2　云架构

就服务类型本身而言，AWS 主要提供六类服务：计算和网络、存储和内容分发、数据库、分析、部署管理以及应用服务。

OpenStack 作为 AWS 的跟随者，它的体系结构里不可避免地体现着 AWS 的各个组件的痕迹。OpenStack 模块设计如图 1-3 所示。

OpenStack 目前的七个核心组件，分别是计算（Compute）、对象存储（Object Storage）、认证（Identity）、用户界面（Dashboard）、块存储（Block Storage）、网络（Network）和镜像服务（Image Service）。

Compute 的项目代号是 Nova，它根据需求提供虚拟机服务，如创建虚拟机或对虚拟机做热迁移。

Object Storage 的项目代号是 Swift，它允许存储或检索对象，也可以认为它允许存储或检索文件。

Identity 的项目代号是 Keystone，为所有 OpenStack 服务提供身份验证和授权，跟踪用户及他们的权限。

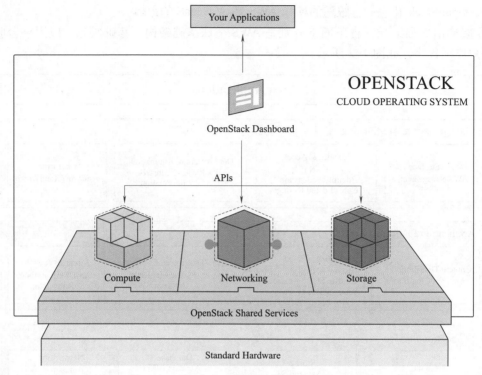

图 1-3　模型设计展示

Dashbord 的项目代号是 Horizon，它为所有 OpenStack 的服务提供一个模块化的基于 Django 的界面。

Block Storage 的项目代号是 Cinder，提供块存储服务。

Network 的项目代号是 Neutron，用于提供网络连接服务，允许用户创建自己的虚拟网络并连接各种网络设备接口。

Image Service 的项目代号是 Glance，它是 OpenStack 的镜像服务组件。

1.6.2　OpenStack 核心项目

OpenStack 覆盖网络、虚拟化、操作系统、服务器等各方面。它是一个正在开发中的云计算平台项目，根据成熟及重要程度的不同，被分解成核心项目、孵化项目，以及支持项目和相关项目。每个项目都有其委员会和项目技术主管，而且每个项目都不是一成不变的，孵化项目可以根据发展的成熟度和重要性，转变为核心项目。截至 Icehouse 版本，下面列出了 10 个核心项目（即 OpenStack 服务）。

（1）计算（Compute）：Nova。一套控制器，用于为单个用户或使用群组管理虚拟机实例的整个生命周期，根据用户需求来提供虚拟服务。负责虚拟机创建、开机、关机、挂起、暂停、调整、迁移、重启、销毁等操作，配置 CPU、内存等信息规格。自 Austin 版本集成到项目中。

（2）对象存储（Object Storage）：Swift。一套用于在大规模可扩展系统中通过内置冗余及高容错机制实现对象存储的系统，允许进行存储或者检索文件。可为 Glance 提供镜像存储，为 Cinder 提供卷备份服务。自 Austin 版本集成到项目中。

（3）镜像服务（Image Service）：Glance。一套虚拟机镜像查找及检索系统，支持多种虚拟机镜像格式（AKI、AMI、ARI、ISO、QCOW2、Raw、VDI、VHD、VMDK），有创建上传镜像、删除镜像、编辑镜像基本信息的功能。自 Bexar 版本集成到项目中。

（4）身份服务（Identity Service）：Keystone。为 OpenStack 其他服务提供身份验证、服务规则和服务令牌的功能，管理 Domains、Projects、Users、Groups、Roles。自 Essex 版本集成到项目中。

（5）网络 & 地址管理（Network）：Neutron。提供云计算的网络虚拟化技术，为 OpenStack 其他服务提供网络连接服务。为用户提供接口，可以定义 Network、Subnet、Router，配置 DHCP、DNS、负载均衡、L3 服务，网络支持 GRE、VLAN。插件架构支持许多主流的网络厂家和技术，如 OpenvSwitch。自 Folsom 版本集成到项目中。

（6）块存储（Block Storage）：Cinder。为运行实例提供稳定的数据块存储服务，它的插件驱动架构有利于块设备的创建和管理，如创建卷、删除卷，在实例上挂载和卸载卷。自 Folsom 版本集成到项目中。

（7）UI 界面（Dashboard）：Horizon。OpenStack 中各种服务的 Web 管理门户，用于简化用户对服务的操作，如启动实例、分配 IP 地址、配置访问控制等。自 Essex 版本集成到项目中。

（8）测量（Metering）：Ceilometer。像一个漏斗一样，能把 OpenStack 内部发生的绝大多数的事件都收集起来，然后为计费和监控以及其他服务提供数据支撑。自 Havana 版本集成到项目中。

（9）部署编排（Orchestration）：Heat[2]。提供了一种通过模板定义的协同部署方式，实现云基础设施软件运行环境（计算、存储和网络资源）的自动化部署。自 Havana 版本集成到项目中。

（10）数据库服务（Database Service）：Trove。为用户在 OpenStack 的环境提供可扩展和可靠的关系和非关系数据库引擎服务。自 Icehouse 版本集成到项目中。

1.6.3　OpenStack 典型环境架构

OpenStack 既是一个社区，也是一个项目和一个开源软件，提供开放源码软件，建立公有云和私有云，它提供了一个部署云的操作平台或工具集，其宗旨在于：帮助组织运行为虚拟计算或存储服务的云，既为公有云、私有云，也为大云、小云提供可扩展的、灵活的云计算。

OpenStack 开源项目由社区维护，包括 OpenStack 计算（代号为 Nova）、OpenStack 对象存储（代号为 Swift），以及 OpenStack 镜像服务（代号 Glance）的集合。OpenStack 提供了一个操作平台或工具包，用于编排云。

Openstack 的详细构架如图 1-4 和图 1-5 所示。

整个 OpenStack 是由控制结点、计算结点、网络结点和存储结点四大部分组成。（这四个结点也可以安装在一台机器上，单机部署）。

其中，控制结点负责对其余结点的控制，包含虚拟机建立、迁移、网络分配、存储分配等；计算结点负责虚拟机运行；网络结点负责对外网络与内网络之间的通信；存储结点负责对虚拟机的额外存储管理等。

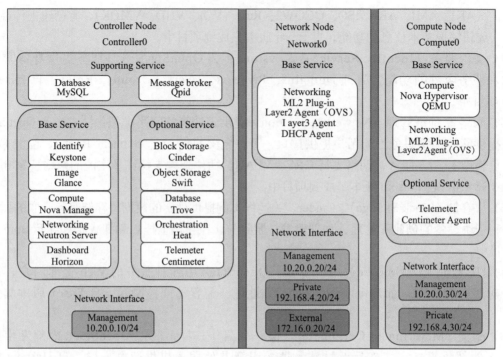

图 1-4　OpenStack 的网络拓扑结构图 1

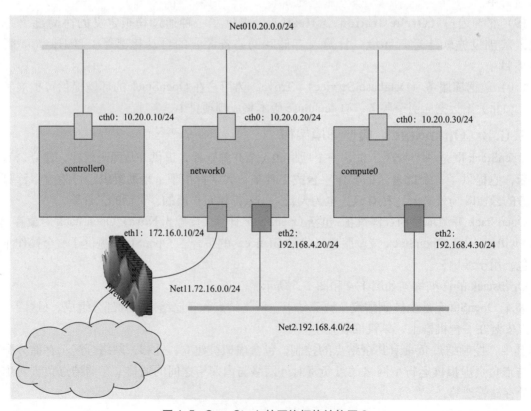

图 1-5　OpenStack 的网络拓扑结构图 2

1. 控制结点架构

控制结点包括以下服务：管理支持服务、基础管理服务、扩展管理服务。

1）管理支持服务

管理支持服务包含 MySQL 与 Qpid 两个服务。

MySQL：数据库作为基础/扩展服务产生的数据存放的地方。

Qpid：消息代理（又称消息中间件）为其他各种服务之间提供统一的消息通信服务。

2）基础管理服务

基础管理服务包含 Keystone、Glance、Nova、Neutron、Horizon 五个服务。

Keystone：认证管理服务，提供其余所有组件的认证信息/令牌的管理、创建、修改等，使用 MySQL 作为统一的数据库。

Glance：镜像管理服务，提供对虚拟机部署时所能提供的镜像的管理，包含镜像的导入、格式，以及制作相应的模板。

Nova：计算管理服务，提供对计算结点的 Nova 的管理，使用 Nova-API 进行通信。

Neutron：网络管理服务，提供对网络结点的网络拓扑管理，同时提供 Neutron 在 Horizon 的管理面板。

Horizon：控制台服务，提供以 Web 的形式对所有结点的所有服务的管理，通常把该服务称为 DashBoard。

3）扩展管理服务

扩展管理服务包含 Cinder、Swift、Trove、Heat、Centimeter 五个服务。

Cinder：提供管理存储结点的 Cinder 相关，同时提供 Cinder 在 Horizon 中的管理面板。

Swift：提供管理存储结点的 Swift 相关，同时提供 Swift 在 Horizon 中的管理面板。

Trove：提供管理数据库结点的 Trove 相关，同时提供 Trove 在 Horizon 中的管理面板。

Heat：提供基于模板来实现云环境中资源的初始化，依赖关系处理，部署等基本操作，也可以解决自动收缩、负载均衡等高级特性。

Centimeter：提供对物理资源以及虚拟资源的监控，并记录这些数据，对该数据进行分析，在一定条件下触发相应动作。

控制结点一般来说只需要一个网络端口用于通信/管理各个结点。

2. 网络结点架构

网络结点仅包含 Neutron 服务。

Neutron：负责管理私有网段与公有网段的通信，以及管理虚拟机网络之间的通信/拓扑，管理虚拟机之上的防火等。

网络结点包含三个网络端口：

eth0：用于与控制结点进行通信。

eth1：用于与除了控制结点之外的计算/存储结点之间的通信。

eth2：用于外部的虚拟机与相应网络之间的通信。

3. 计算结点架构

计算结点包含 Nova、Neutron、Telemeter 三个服务。

1）基础服务

Nova：提供虚拟机的创建、运行、迁移、快照等各种围绕虚拟机的服务，并提供 API 与控制结点对接，由控制结点下发任务。

Neutron：提供计算结点与网络结点之间的通信服务。

2）扩展服务

Telmeter：提供计算结点的监控代理，将虚拟机的情况反馈给控制结点，是 Centimeter 的代理服务。

计算结点包含最少两个网络端口：

eth0：与控制结点进行通信，受控制结点统一调配。

eth1：与网络结点、存储结点进行通信。

4. 存储结点架构

存储结点包含 Cinder、Swift 等服务。

Cinder：块存储服务，提供相应的块存储，简单来说，就是虚拟出一块磁盘，可以挂载到相应的虚拟机之上，不受文件系统等因素影响，对虚拟机来说，这个操作就像是新加一块硬盘，可以完成对磁盘的任何操作，包括挂载、卸载、格式化、转换文件系统等操作，大多应用于虚拟机空间不足的情况下的空间扩容等。

Swift：对象存储服务，提供相应的对象存储，简单来说，就是虚拟出一块磁盘空间，可以在这个空间当中存放文件，也仅仅只能存放文件，不能进行格式化、转换文件系统等操作，大多应用于云磁盘 / 文件。

存储结点包含最少两个网络接口：

eth0：与控制结点进行通信，接受控制结点任务，受控制结点统一调配。

eth1：与计算 / 网络结点进行通信，完成控制结点下发的各类任务。

单元 2

CentOS 基本环境配置

学习目标

◎ 掌握在虚拟机环境中部署安装 CentOS 系统；
◎ 掌握 Linux 常用命令，镜像挂载；
◎ 掌握使用 yum 源安装软件包，并能进行网络规划、网卡配置和故障排查。

2.1 安装 CentOS7 操作系统

①打开 VMware Workstation 虚拟机，选择"创建新的虚拟机"，如图 2-1 所示。

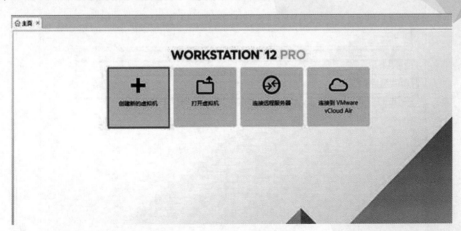

图 2-1 创建虚拟机

②选择"Linux"→"CentOS 64 位"操作系统,如图 2-2 所示。

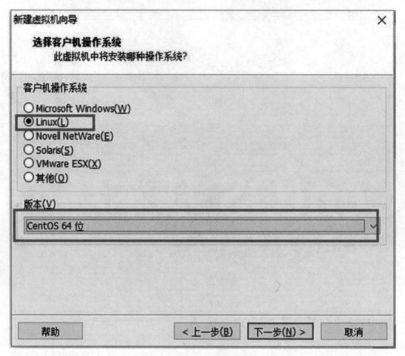

图 2-2 选择操作系统

③为虚拟机命名,并选择安装位置,如图 2-3 所示。

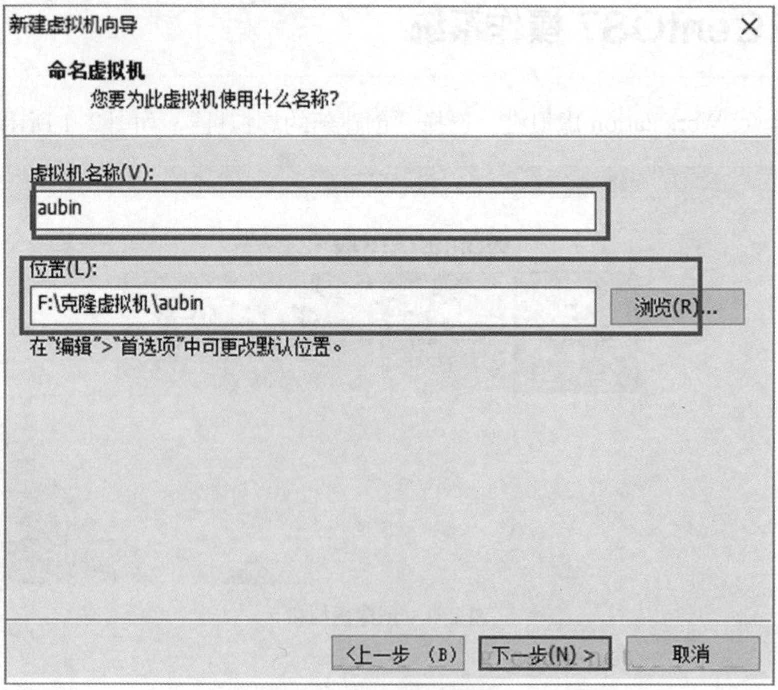

图 2-3 选择安装位置

④设定虚拟机存储方式，并设定磁盘大小，如图 2-4 所示。

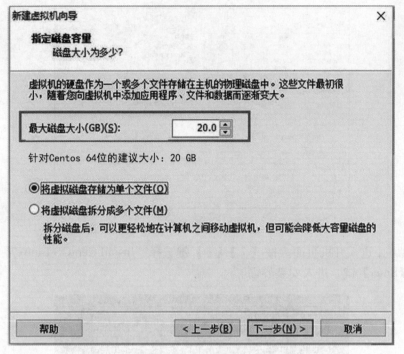

图 2-4　设定磁盘大小

⑤完成虚拟机的创建，如图 2-5 所示。

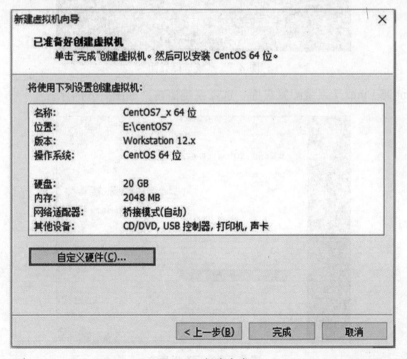

图 2-5　创建完成

⑥导入 CentOS Linux7 系统镜像，如图 2-6 所示。

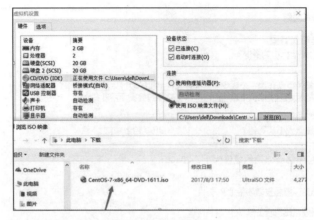

图 2-6　导入镜像

⑦打开虚拟机，进入开机页面。按【↑】【↓】键选择"Install CentOS Linux7"选项，如图 2-7 所示。然后按【Enter】键，进入安装界面。

图 2-7　系统选择界面

⑧进入 CentOS Linux 7 系统配置页面，选择系统语言，如图 2-8 所示。

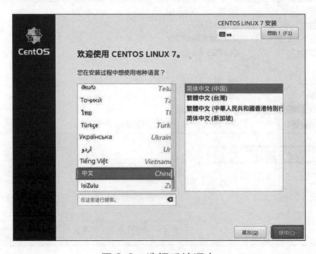

图 2-8　选择系统语言

⑨安装信息摘要，如图 2-9 所示。

图 2-9　安装信息摘要

⑩软件选择：选择"最小安装"，如图 2-10 所示。

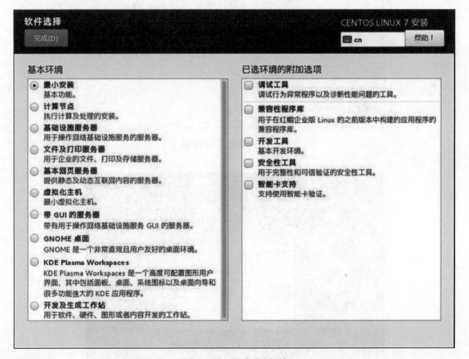

图 2-10　最小化安装

⑪对硬盘进行配置，首先选择安装位置，再选择手动进行分区。
添加挂载点"/boot"分区，并挂载一个"/swap"分区和一个根分区。操作步骤如图 2-11~图 2-16 所示。

图 2-11　硬盘配置分区 1

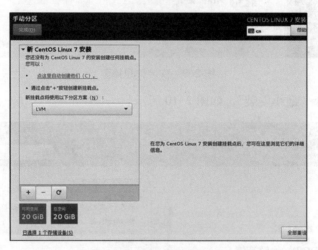

图 2-12　硬盘配置分区 2

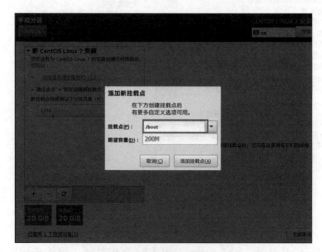

图 2-13　硬盘配置分区 3

单元 2　CentOS 基本环境配置

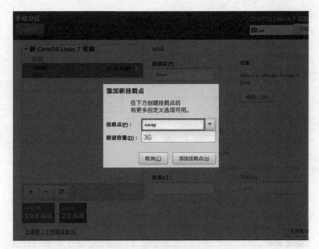

图 2-14　硬盘配置分区 4

图 2-15　硬盘配置分区 5

图 2-16　硬盘配置分区 6

⑫配置完成后，开始安装操作系统，如图 2-17 所示。

图 2-17　安装操作系统

⑬进行 ROOT 密码的设置和用户的创建，如图 2-18 所示。

图 2-18　密码设置和用户创建

⑭安装完成后，重启，如图 2-19 所示。

图 2-19　安装完成后重启

⑮CentOS Linux 7操作系统安装完成，显示系统信息，输入用户名密码方可使用，如图2-20所示。

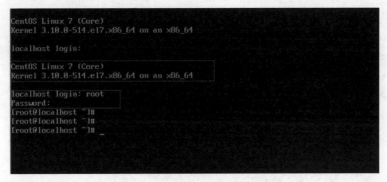

图 2-20　安装完成显示系统信息

2.2　Linux 基本技能

2.2.1　常用命令

①文件与目录操作，命令如图 2-21 所示。

命令	解析
cd /home	进入 '/home' 目录
cd ..	返回上一级目录
cd ../..	返回上两级目录
cd -	返回上次所在目录
cp file1 file2	将file1复制为file2
cp -a dir1 dir2	复制一个目录
cp -a /tmp/dir1 .	复制一个目录到当前工作目录（.代表当前目录）
ls	查看目录中的文件
ls -a	显示隐藏文件
ls -l	显示详细信息
ls -lrt	按时间显示文件（l表示详细列表，r表示反向排序，t表示按时间排序）
pwd	显示工作路径
mkdir dir1	创建 'dir1' 目录
mkdir dir1 dir2	同时创建两个目录
mkdir -p /tmp/dir1/dir2	创建一个目录树
mv dir1 dir2	移动/重命名一个目录
rm -f file1	删除 'file1'
rm -rf dir1	删除 'dir1' 目录及其子目录内容

图 2-21　文件与目录操作命令

②查看文件内容，命令如图2-22所示。

命令	解析
cat file1	从第一个字节开始正向查看文件的内容
head -2 file1	查看一个文件的前两行
more file1	查看一个长文件的内容
tac file1	从最后一行开始反向查看一个文件的内容
tail -3 file1	查看一个文件的最后三行
vi file	打开并浏览文件

图2-22 查看文件内容命令

③文本内容处理，命令如图2-23所示。

命令	解析		
grep str /tmp/test	在文件 '/tmp/test' 中查找 "str"		
grep ^str /tmp/test	在文件 '/tmp/test' 中查找以 "str" 开始的行		
grep [0-9] /tmp/test	查找 '/tmp/test' 文件中所有包含数字的行		
grep str -r /tmp/*	在目录 '/tmp' 及其子目录中查找 "str"		
diff file1 file2	找出两个文件的不同处		
sdiff file1 file2	以对比的方式显示两个文件的不同		
vi file	操作	解析	
	i	进入编辑文本模式	
	Esc	退出编辑文本模式	
	:w	保存当前修改	
	:q	不保存退出vi	
	:wq	保存当前修改并退出vi	

图2-23 文本处理命令

④查询操作，命令如图2-24所示。

命令	解析
find / -name file1	从 '/' 开始进入根文件系统查找文件和目录
find / -user user1	查找属于用户 'user1' 的文件和目录
find /home/user1 -name *.bin	在目录 '/ home/user1' 中查找以 '.bin' 结尾的文件
find /usr/bin -type f -atime +100	查找在过去100天内未被使用过的执行文件
find /usr/bin -type f -mtime -10	查找在10天内被创建或者修改过的文件
locate *.ps	寻找以 '.ps' 结尾的文件，先运行 'updatedb' 命令
find -name "*.[ch]'	xargs grep -E 'expr'	在当前目录及其子目录所有.c和.h文件中查找 'expr'
find -type f -print0	xargs -r0 grep -F 'expr'	在当前目录及其子目录的常规文件中查找 'expr'
find -maxdepth 1 -type f	xargs grep -F 'expr'	在当前目录中查找 'expr'

图2-24 查询操作命令

单元 2　CentOS 基本环境配置

⑤压缩、解压，命令如图 2-25 所示。

命令	解析
bzip2 file1	压缩 file1
bunzip2 file1.bz2	解压 file1.bz2
gzip file1	压缩 file1
gzip -9 file1	最大程度压缩 file1
gunzip file1.gz	解压 file1.gz
tar -cvf archive.tar file1	把 file1 打包成 archive.tar（-c: 建立压缩档案；-v: 显示所有过程；-f: 使用档案名字，是必须的，是最后一个参数）
tar -cvf archive.tar file1 dir1	把 file1，dir1 打包成 archive.tar
tar -tf archive.tar	显示一个包中的内容
tar -xvf archive.tar	释放一个包
tar -xvf archive.tar -C /tmp	把压缩包释放到 /tmp 目录下
zip file1.zip file1	创建一个 zip 格式的压缩包
zip -r file1.zip file1 dir1	把文件和目录压缩成一个 zip 格式的压缩包
unzip file1.zip	解压一个 zip 格式的压缩包到当前目录
unzip test.zip -d /tmp/	解压一个 zip 格式的压缩包到 /tmp 目录

图 2-25　压缩、解压命令

⑥ yum 安装器，命令如图 2-26 所示。

命令	解析
yum -y install [package]	下载并安装一个 rpm 包
yum localinstall [package.rpm]	安装一个 rpm 包，使用你自己的软件仓库解决所有依赖关系
yum -y update	更新当前系统中安装的所有 rpm 包
yum update [package]	更新一个 rpm 包
yum remove [package]	删除一个 rpm 包
yum list	列出当前系统中安装的所有包
yum search [package]	在 rpm 仓库中搜寻软件包
yum clean [package]	清除缓存目录（/var/cache/yum）下的软件包
yum clean headers	删除所有头文件
yum clean all	删除所有缓存的包和头文件

图 2-26　安装器命令

⑦网络相关，命令如图 2-27 所示。

命令	解析
ifconfig eth0	显示一个以太网卡的配置
ifconfig eth0 192.168.1.1 netmask 255.255.255.0	配置网卡的 IP 地址
ifdown eth0	禁用 'eth0' 网络设备
ifup eth0	启用 'eth0' 网络设备
iwconfig eth1	显示一个无线网卡的配置
iwlist scan	显示无线网络
ip addr show	显示网卡的 IP 地址

图 2-27　网络相关命令

⑧系统相关命令如图 2-28 所示。

命令	解析
su -	切换到root权限（与su有区别）
shutdown -h now	关机
shutdown -r now	重启
top	罗列使用CPU资源最多的linux任务（输入q退出）
pstree	以树状图显示程序
man ping	查看参考手册（例如ping 命令）
passwd	修改密码
df -h	显示磁盘的使用情况
cal -3	显示前一个月，当前月以及下一个月的月历
cal 10 1988	显示指定月，年的月历
date --date '1970-01-01 UTC 1427888888 seconds'	把一相对于1970-01-01 00:00的秒数转换成时间

图 2-28 系统相关命令

2.2.2 Vim 编辑器的使用

Vim 的四种模式及转换（见图 2-29）：

（1）命令模式：控制屏幕的光标移动，进行文本的删除、复制等文字编辑工作，当使用 vim 打开某个文件时，默认模式就是命令模式。

（2）插入模式：只有在插入模式下，才可以输入文字。

（3）末行模式：保存文件或退出 vim，同时也可以设置编辑环境和一些编译工作，如列出行号、寻找字符串等。

（4）可视化模式：可以使用鼠标框选文字，比较人性化。

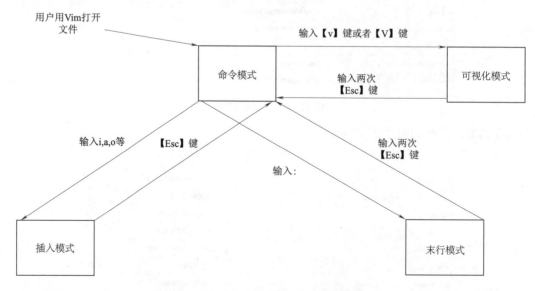

图 2-29 Vim 的四种模式及转换图示

1. 命令模式

[root@localhost ~]# vim filename

在命令模式下可以直接用【Del】键删除字符；在命令模式下可以按【w/q】组合键保存退出，其中【w】键表示保存（write），【q】键表示退出（quit）。

2. 插入模式

如果想要和 Windows 下 txt 文件一样插入命令，需要进入插入模式：

i 表示在光标所在字符前插入

a 表示在光标所在字符后插入

o 表示在光标下插入新行

3. 编辑模式

要先在命令模式中，才能进入编辑模式，进入编辑模式的标志就是输入冒号（：）。

在编辑模式下可以输入 set nu 指令设置行号。常用指令如下：

指令	说明
: set nu	设置行号
: set nonu	取消行号
gg	到第一行
G	到最后一行
: n	到第 n 行
$	移至行尾
0	移至行首
x	删除字符
nx	删除光标所在处多个字符
dd	删除一行
ndd	删除 n 行
yy	复制当前行
nyy	复制当前以下 n 行
p	粘贴到当前光标所在行下
u	取消上一步操作
/string	指定搜索字符串
: w	保存修改
: wq	保存修改并退出
: q!	强制退出不保存
: wq!	强制保存退出

2.2.3 镜像的挂载方式

镜像的挂载步骤如下：

（1）上传镜像文件到服务器的某一目录下（/centos-iso）。

（2）创建镜像文件挂载目录：

mkdir -p /mnt/yum

(3) 挂载镜像：
mount -o loop /centos-iso/centos-7iso /mnt/yum/
* 如果要卸载执行 umount /mnt/yum
如果要实现开机自动挂载，编辑 /etc/fstab，并添加一行：
/centos-iso/centos-7iso /mnt/yum iso9660 loop 0 0

2.2.4 Yum 源的配置及软件包的安装

Yum 软件仓库的作用是为了进一步简化 RPM 管理软件的难度以及自动分析所需软件包及其依赖关系的技术。可以把 Yum 想象成一个硕大的软件仓库，里面保存绝大部分的常用工具，而且只需要说出所需的软件包名称，系统就会自动为用户搞定一切。搭建并配置 Yum 软件仓库的步骤如下：

第 1 步：进入 /etc/yum.repos.d/ 目录（因为该目录存放着 Yum 软件仓库的配置文件）。

第 2 步：使用 Vim 编辑器创建一个名为 rhel7.repo 的新配置文件（文件名可随意，但扩展名必须为 .repo），逐项写入下面加粗的配置参数并保存退出（不要写后面的中文注释）。

Ø [rhel-media]：Yum 软件仓库唯一标识符，避免与其他仓库冲突。
Ø name=wz：Yum 软件仓库的名称描述，易于识别仓库用处。
Ø baseurl=file:///media/cdrom：提供的方式包括 FTP（ftp://..）、HTTP（http://..）、本地（file:///..）。
Ø enabled=1：设置此源是否可用；1 为可用，0 为禁用。
Ø gpgcheck=1：设置此源是否校验文件；1 为校验，0 为不校验。
Ø gpgkey=file:///media/cdrom/RPM-GPG-KEY-redhat-release：若上面参数开启校验，那么请指定公钥文件地址。

第 3 步：按配置参数的路径挂载光盘，并把光盘挂载信息写入 /etc/fstab 文件。

第 4 步：使用 "yum install httpd-y" 命令检查 Yum 软件仓库是否已经可用。

进入 /etc/yum.repos.d 目录后创建 Yum 配置文件：

```
[root@wz ~]# cd /etc/yum.repos.d/
[root@wz yum.repos.d]# vim rhel7.repo
[rhel7]
name=rhel7
baseurl=file:///media/cdrom
enabled=1
gpgcheck=0
```

创建挂载点后进行挂载操作，并设置为开机自动挂载。尝试使用 Yum 软件仓库来安装 Web 服务，出现 "Complete！" 则代表配置正确：

```
[root@wz yum.repos.d]# mkdir -p /media/cdrom
[root@wz yum.repos.d]# mount /dev/cdrom /media/cdrom
mount: /dev/sr0 is write-protected, mounting read-only
[root@wz yum.repos.d]# vim /etc/fstab
/dev/cdrom /media/cdrom iso9660 defaults 0 0
[root@wz ~]# yum install httpd
```

```
Loaded plugins: langpacks, product-id, subscription-manager
……………省略部分输出信息……………
Dependencies Resolved
================================================================================
 Package          Arch      Version             Repository   Size
================================================================================
Installing:
 httpd            x86_64    2.4.6-17.el7        rhel         1.2 M
Installing for dependencies:
 apr              x86_64    1.4.8-3.el7         rhel         103 k
 apr-util         x86_64    1.5.2-6.el7         rhel         92 k
 httpd-tools      x86_64    2.4.6-17.el7        rhel         77 k
 mailcap          noarch    2.1.41-2.el7        rhel         31 k
Transaction Summary
================================================================================
Install 1 Package (+4 Dependent packages)
Total download size: 1.5 M
Installed size: 4.3 M
Is this ok [ẏ/d/N]: y
Downloading packages:
--------------------------------------------------------------------------------
……………省略部分输出信息……………
Complete!
```

2.3 网络配置

2.3.1 网卡文件的配置

网卡 IP 地址的配置是否正确是两台服务器是否可以相互通信的前提。在 Linux 系统中，一切都是文件，因此配置网络服务的工作其实就是在编辑网卡配置文件。

在 CentOS 7 中，网卡配置文件的前缀则以 ifcfg 开始，加上网卡名称共同组成网卡配置文件的名字，如 ifcfg-eno16777736。

现在有一个名称为 ifcfg-eno16777736 的网卡设备，将其配置为开机自启动，并且 IP 地址、子网、网关等信息由人工指定，其步骤如下：

第 1 步：首先切换到 /etc/sysconfig/network-scripts 目录中（存放着网卡的配置文件）。

第 2 步：使用 Vim 编辑器修改网卡文件 ifcfg-eno16777736，逐项写入下面的配置参数并保存退出。由于每台设备的硬件及架构是不一样的，因此请读者使用 ifconfig 命令自行确认各自网卡的默认名称。

- Ø 设备类型：TYPE=Ethernet
- Ø 地址分配模式：BOOTPROTO=static
- Ø 网卡名称：NAME=eno16777736
- Ø 是否启动：ONBOOT=yes
- Ø IP 地址：IPADDR=192.168.10.10
- Ø 子网掩码：NETMASK=255.255.255.0
- Ø 网关地址：GATEWAY=192.168.10.1
- Ø DNS 地址：DNS1=192.168.10.1

第 3 步：重启网络服务并测试网络是否连通。

进入网卡配置文件所在的目录，然后编辑网卡配置文件，在其中填入下面的信息：

```
[root@wz ~]# cd /etc/sysconfig/network-scripts/
[root@wz network-scripts]# vim ifcfg-eno16777736
TYPE=Ethernet
BOOTPROTO=static
NAME=eno16777736
ONBOOT=yes
IPADDR=192.168.10.10
NETMASK=255.255.255.0
GATEWAY=192.168.10.1
DNS1=192.168.10.1
```

执行重启网卡设备的命令（在正常情况下不会有提示信息），然后通过 ping 命令测试网络能否连通。由于在 Linux 系统中 ping 命令不会自动终止，因此需要手动按下【Ctrl+C】组合键来强行结束进程。

```
[root@wz network-scripts]# systemctl restart network
[root@wz network-scripts]# ping 192.168.10.10
PING 192.168.10.10 (192.168.10.10) 56(84) bytes of data.
64 bytes from 192.168.10.10: icmp_seq=1 ttl=64 time=0.081 ms
64 bytes from 192.168.10.10: icmp_seq=2 ttl=64 time=0.083 ms
64 bytes from 192.168.10.10: icmp_seq=3 ttl=64 time=0.059 ms
64 bytes from 192.168.10.10: icmp_seq=4 ttl=64 time=0.097 ms
^C
--- 192.168.10.10 ping statistics ---
4 packets transmitted, 4 received, 0% packet loss, time 2999ms
rtt min/avg/max/mdev=0.059/0.080/0.097/0.013 ms
```

2.3.2 解决常见网络故障

用户有时更改完静态 IP 后，会发现 Network 服务重启不了，此时可执行 systemctl start network 命令查看系统报错原因。

```
$ systemctl start network
Job for network.service failed because the control process exited with error code. See "systemctl status network.service" and "journalctl -xe" for details.
```

根据提示输入 systemctl status network 命令后出现如下错误信息：

```
$  systemctl status network
● network.service - LSB: Bring up/down networking
  Loaded: loaded (/etc/rc.d/init.d/network; bad; vendor preset: disabled)
  Active: failed (Result: exit-code) since Tue 2018-10-09 22:47:07 CST; 2min 5s ago
  Docs: man:systemd-sysv-generator(8)
  Process: 8980 ExecStart=/etc/rc.d/init.d/network start (code=exited, status=1/FAILURE)

Oct 09 22:47:07 Server01 network[8980]: RTNETLINK answers: File exists
Oct 09 22:47:07 Server01 network[8980]: RTNETLINK answers: File exists
Oct 09 22:47:07 Server01 network[8980]: RTNETLINK answers: File exists
Oct 09 22:47:07 Server01 network[8980]: RTNETLINK answers: File exists
Oct 09 22:47:07 Server01 network[8980]: RTNETLINK answers: File exists
Oct 09 22:47:07 Server01 network[8980]: RTNETLINK answers: File exists
Oct 09 22:47:07 Server01 systemd[1]: network.service: control process exited, code=exited status=1
Oct 09 22:47:07 Server01 systemd[1]: Failed to start LSB: Bring up/down networking.
Oct 09 22:47:07 Server01 systemd[1]: Unit network.service entered failed state.
Oct 09 22:47:07 Server01 systemd[1]: network.service failed.
```

Network 服务无法启动时，首先保证 /etc/sysconfig/network-scripts 目录下的 ifcfg-xxx 没有错误，若不是此类错误，则按下述步骤执行：

（1）和 NetworkManager 服务有冲突时，直接关闭 NetworkManger 服务即可，命令为 systemctl stop NetworkManager，并且禁止开机启动 chkconfig NetworkManager off。之后重启即可。

（2）和配置文件的 MAC 地址不匹配时，使用 ip addr（或 ifconfig）查看 ens33 下的 MAC 地址 00：0c：29：b1：44：a0，将 /etc/sysconfig/network-scripts/ifcfg-xxx 中的 HWADDR（如果没有就添加上）改成这个 MAC 地址。

```
$ ip addr
1: lo: <LOOPBACK,UP,LOWER_UP> mtu 65536 qdisc noqueue state UNKNOWN
    link/loopback 00:00:00:00:00:00 brd 00:00:00:00:00:00
    inet 127.0.0.1/8 scope host lo
       valid_lft forever preferred_lft forever
    inet6 ::1/128 scope host
       valid_lft forever preferred_lft forever
2: ens33: <BROADCAST,MULTICAST,UP,LOWER_UP> mtu 1500 qdisc pfifo_fast state UP qlen 1000
    link/ether 00:0c:29:b1:44:a0 brd ff:ff:ff:ff:ff:ff
    inet 192.168.1.102/24 brd 192.168.1.255 scope global ens33
       valid_lft forever preferred_lft forever
    inet6 fe80::20c:29ff:feb1:44a0/64 scope link
       valid_lft forever preferred_lft forever
3: br-3097ed36fd04: <NO-CARRIER,BROADCAST,MULTICAST,UP> mtu 1500 qdisc noqueue state DOWN
    link/ether 02:42:e3:f2:63:74 brd ff:ff:ff:ff:ff:ff
    inet 172.25.0.1/16 brd 172.25.255.255 scope global br-3097ed36fd04
       valid_lft forever preferred_lft forever
4: br-4d153d29100f: <NO-CARRIER,BROADCAST,MULTICAST,UP> mtu 1500 qdisc noqueue state DOWN
    link/ether 02:42:29:e8:35:29 brd ff:ff:ff:ff:ff:ff
    inet 172.19.0.1/16 brd 172.19.255.255 scope global br-4d153d29100f
       valid_lft forever preferred_lft forever
    inet6 fe80::42:29ff:fee8:3529/64 scope link
       valid_lft forever preferred_lft forever
```

（3）设定开机启动一个名为 NetworkManager-wait-online 服务，命令为：
systemctl enable NetworkManager-wait-online.service

（4）查看 /etc/sysconfig/network-scripts 目录下的 ifcfg-xxx 文件，xxx 需要和使用 ip addr 命令查看 ip 的第二条开头的名称一致，若不一致则需要用 mv 命令修改文件名。

（5）将 ifcfg-xxx 文件中的 DEVICE 和 NAME 都改成 xxx。

单元 3
MySQL 数据库中数据的基本操作

学习目标

◎ 掌握虚拟机环境中部署安装 MySQL 数据库；
◎ 掌握 MySQL 常用的查询、添加、删除命令；
◎ 掌握 MySQL 数据库的用户管理和权限管理。

3.1 在 CentOS 中 MySQL 数据库安装及操作

3.1.1 MySQL 数据库的安装

CentOS 中默认安装有 MariaDB，在安装 MySQL 时会覆盖 MariaDB。这里采用 Yum 管理各种 RPM 包的依赖，能够从指定的服务器自动下载 RPM 包并且安装，但是在安装完成后必须卸载，否则会自动更新。（所有的操作都是切换到 root 用户下进行。）

1. 安装 MySQL 官方的 Yum Repository

[root@localhost ~]# wget -i -c http://dev.mysql.com/get/mysql57-community-release-el7-10.noarch.rpm

2. 下载 RPM 包

[root@localhost ~]# yum -y install mysql57-community-release-el7-10.noarch.rpm

3. 安装 MySQL 服务

[root@localhost ~]# yum -y install mysql-community-server

服务安装的时间比较长，耐心等待，中间有一次询问 y/n？输入 y 后按【Enter】键。

4. 启动 MySQL 服务

[root@localhost ~]# systemctl start mysqld.service

最后两行命令出现则代表启动成功。
Starting MySQL Server...
Started MySQL Server.

5. 登录用户

[root@localhost ~]# mysql -u root -p

首次登录需要输入生成的初始密码，登录后必须要修改这个密码。

初始密码默认是 /var/log/mysqld.log，找到下面这行命令，临时密码是：h.zdWmt/l0M3。

2018-08-22T07:06:29.387527Z 1 [Note] A temporary password is generated for root@localhost: h.zdWmt/l0M3

【注意】

只有启动过一次 MySQL 才可以查看临时密码。

6. 重新设置密码

重新设置密码之前，需要根据用户的需求设置密码的复杂度和长度时，要登录数据库后设置如下参数：

(1) validate_password_policy 表示密码策略，默认是 1：符合长度，且必须含有数字、小写或大写字母、特殊字符。设置为 0 判断密码的标准基于密码的长度。一定要先修改两个参数再修改密码。命令如下：

mysql> set global validate_password_policy=0;

(2) validate_password_length 表示密码长度，最小值为 4。命令如下：

mysql> set global validate_password_length=4;

通过命令查看设置参数的情况：

mysql> SHOW VARIABLES LIKE'validate_password%';
+--------------------------------------+-------+
| Variable_name | Value |
+--------------------------------------+-------+
validate_password_dictionary_file	
validate_password_length	6
validate_password_mixed_case_count	2
validate_password_number_count	1
validate_password_policy	LOW
validate_password_special_char_count	1
+--------------------------------------+-------+
6 rows in set (0.00 sec)

【注意】

以上两行设置都是临时设置，重启数据库后需要重新设置，再创建新密码。修改用户 root 的密码，之后就可以使用该密码登录。

mysql> ALTER USER 'root'@'localhost' IDENTIFIED BY 'mima';
// 代码用户为 root 的密码是 mima

7. 退出数据库

退出数据库命令如下：

```
exit
```
【注意】
① /etc/my.cnf——mysql 的主配置文件。
② /var/lib/mysql——mysql 数据库的数据库文件存放位置。
③ /var/log mysql——数据库的日志输出存放位置。

3.1.2 MySQL 数据库的备份与还原

1. MySQL 数据备份

在平台上新建虚拟机后安装 MySQL，安装完成后 mysqldump 文件在 /usr/bin/mysqldump 目录下。在 /usr/local/mysql/bin 文件夹中有一个 mysqldump 文件，可以进入查看：

```
[root@localhost ~]#cd /usr/local/mysql/
[root@localhost ~]#ls
```

数据库中有一个 pytest 数据库，备份这个数据库到 /home/ 目录下：

```
[root@localhost ~]#/usr/local/mysql/bin/mysqldump -u root -p pytest > /home/pytest.sql
```

当在 /home/ 目录下看到 pytest.sql 文件，就表示备份成功。

2. MySQL 数据恢复

在恢复数据之前，需要建立一个新的数据库，建立的数据库名称是 test1。命令如下：

```
[root@localhost ~]#mysql -u root -p test1 < /home/pytest.sql
```

进入数据库可以查看有没有数据。

如果要备份某一张表，在数据库名加". 表名"，如"test1.student"。

如果备份多个数据库，用","隔开。

3.1.3 MySQL 数据库用户管理

1. 创建用户：以 root 用户登录数据库进行用户创建

创建用户命令如下：

```
CREATE USER 'username'@'host' IDENTIFIED BY 'password';
```

例如，其他用户的创建：

```
CREATE USER 'test_admin'@'localhost' IDENTIFIED BY 'admin@123_S';
CREATE USER 'test_admin2'@'%' IDENTIFIED BY '';
```

【注意】
① username——将要创建的用户名。
② host——指定该用户在哪个主机上可以登录，"localhost"指该用户只能在本地登录，不能在另外一台计算机上远程登录，如果想远程登录，将"localhost"改为"%"，表示在任何一台计算机上都可以登录；也可以指定某台计算机可以远程登录。
③ password——该用户的登录密码,密码可以为空,若为空则该用户可以不用密码登录服务器。

2. 删除账户及权限

删除账户及权限命令如下：

```
drop user 'username'@'host';
```

3.1.4 MySQL 数据库的权限管理

1. 授权：以 root 用户登录数据库进行授权

命令如下：

GRANT privileges ON databasename.tablename TO 'username'@'host'

【注意】

① privileges——用户的操作权限，如 SELECT 等。如果授予所有权限则使用 ALL。

② databasename——数据库名称。

③ tablename——表名。如果要给该用户授予对所有数据库和表的相应操作权限则可用*表示。

例如：

GRANT SELECT ON test_db.* TO 'test_admin2'@'%';
flush privileges;

2. 撤销用户权限

命令如下：

REVOKE privilege ON databasename.tablename FROM 'username'@'host';

例如：

REVOKE SELECT ON test_db.* FROM 'test_min'@'%';

3. 查看用户的授权

命令如下：

SHOW GRANTS FOR 'username'@'host'

例如：

```
mysql> SHOW GRANTS FOR 'test_admin'@'%' ;
+------------------------------------------------------------------+
| Grants for test_admin@%                                          |
+------------------------------------------------------------------+
| GRANT USAGE ON *.* TO 'test_admin'@'%'                           |
| GRANT ALL PRIVILEGES ON 'test_manage_db'.* TO 'test_admin'@'%'   |
+------------------------------------------------------------------+
2 rows in set ( 0.00 sec)
```

3.2 在 MySQL 数据库中插入数据

3.2.1 为表中所有字段添加数据

MySQL 使用 INSERT 语句，向数据表中添加数据，根据添加方式的不同，分为以下三种：

（1）为表的所有字段添加数据。

（2）为表的指定字段添加数据。

（3）同时添加多条记录。

一般情况下，向数据表中添加新记录，应该包含表的所有字段，为表的所有字段添加数据，使用 INSERT 语句。命令如下：
　　INSERT INTO 表名（字段名1，字段名2，…）
　　VALUES（值1，值2，…）；
参数说明：
(1) 字段名1,字段名2,……表示数据表中的字段名称。此处,必须列出表中所有字段的名称。
(2) 值1，值2，……表示每个字段的值，每个值的顺序、类型必须与对应的字段相匹配。
1. 创建一个数据库 chapter03
命令如下：
mysql>CREATE DATABASE chapter03;
Query OK, 1 row affected（0.00 sec）
mysql>USE chapter03;
Database changed
2. 创建一个表 student
命令如下：
mysql>CREATE TABLE student（
　　->id INT（4），
　　->name VARCHAR（20）NOT NULL,
　　->grade FLOAT
　　->）;
Query OK, 0 row affected（2.06 sec）
3. 向 student 表中添加一条记录
id 字段的值为1，name 字段的值为 zhangsan，grade 字段的值为98。命令如下：
mysql>INSERT INTO student（id name, grade）
　　->VALUES（1, 'ZHANGSAN', 98）;
Query OK, 1 row affected（1.72 sec）
mysql>SELECT * FROM student;
+------+----------+-------+
| id | name | grade |
| 1 | zhangsan | 98 |
+------+----------+-------+
1 row in set（0.00 sec）

3.2.2　为表中指定字段添加数据

只向表的部分字段添加数据，而其他字段的值为表定义时的默认值。命令如下：
INSERT INTO 表名
SET 字段名1=值1[，字段名2=值2，……]
　　如果在 set 关键字后面，指定多个字段名=值对，每对之间使用逗号分隔，最后一个字段名=值对之后，不需要逗号。
　　例如：向 student 表中添加一条记录，id 字段的值为5，name 字段的值为 boya，grade 字段的值为99。

```
mysql>INSERT INTO student
    ->SET id=5, name='boya', grade=99;
Query OK, 1 row affected(0.08 sec)
```

3.2.3 同时添加多条记录

调用多次 INSERT 语句可以插入多条记录，但使用这种方法会增加服务器的负荷，因为，每一次执行 SQL，服务器都要对 SQL 进行分析、优化等操作。MySQL 提供了使用一条 INSERT 语句来插入多条记录的方法。例如：

```
mysql>insert into persons (id_p, lastname , firstName, city )
    ->values
    ->(200, 'haha' , 'deng' , 'shenzhen'),
    ->(201, 'haha2' , 'deng' , 'GD'),
    ->(202, 'haha3' , 'deng' , 'Beijing');
Query OK, 3 row affected(0.50 sec)
```

3.3 在 MySQL 数据库中更新数据

1. MySQL-UPDATE 语句

用于更新表中的现有数据。亦可用 UPDATE 语句来更改表中单个行、一组行或所有行的列值。

MySQL-UPDATE 语法：

```
UPDATE [LOW_PRIORITY] [IGNORE] table_name
SET
    column_name1=expr1,
    column_name2=expr2,
    ...
WHERE
    condition;
```

在上面的 UPDATE 语句中，需注意以下几点：

第一，在 UPDATE 关键字后面指定要更新数据的表名。

第二，SET 子句指定要修改的列和新值。要更新多个列，请使用以逗号分隔的列表。

第三，使用 WHERE 语句中的条件指定要更新的行。WHERE 子句是可选的。如果省略 WHERE 子句，则 UPDATE 语句将更新表中的所有行。

2. MySQL-UPDATE 多列

更新多列中的值，需要在 SET 子句中指定分配。

例如更新员工编号 1056 的姓氏和电子邮件列的命令如下：

```
UPDATE employees
SET
    lastname='Hill',
    email='mary.hill@yiibai.com'
```

```
WHERE
    employeeNumber=1056;
```

3. MySQL-UPDATE JOIN 语句

在 MySQL 中,可以在 UPDATE 语句中使用 JOIN 子句执行跨表更新。

JOIN 语句的作用:查询表中的行(在 INNER JOIN 的情况下);查询另一个表中的相应行(在 LEFT JOIN 的情况下)。

UPDATE JOIN 语法:

```
UPDATE T1, T2,
[INNER JOIN | LEFT JOIN] T1 ON T1.C1=T2. C1
SET T1.C2=T2.C2,
    T2.C3=expr
WHERE condition
```

UPDATE JOIN 语法说明:

第一,在 UPDATE 子句之后,指定主表(T1)和从表(T2)。

第二,指定一种要使用的连接,即 INNER JOIN 或 LEFT JOIN 和连接条件。JOIN 子句必须出现在 UPDATE 子句之后。

第三,要为要更新的 T1 或 T2 表中的列分配新值。

第四,WHERE 语句中的条件用于指定要更新的行。

更新数据交叉表的另一种方法如下:

```
UPDATE T1, T2
INNER JOIN T2 ON T1.C1=T2.C1
SET T1.C2=T2.C2,
    T2.C3=expr
WHERE condition
```

3.4 在 MySQL 数据库中查询数据

3.4.1 SELECT 语法

SELECT 语法:

```
SELECT COL1, COL2,...COLn FROM TABLE1, TABLE2,...TABLEn
[WHERE CONDITIONS]  -- 查询条件
[GROUP BY GROUP_BY_LIST]  -- 查询结果分组
[HAVING CONDITIONS]  -- 查询条件-统计结果作为条件
[ORDER BY ORDER_LIST[ASC|DESC]]  -- 查询结果排序
```

3.4.2 简单查询

1. 查询表的全部行和列

例如:查询玩家表中全部的行和列。

```
select  user_qq, user_name, user_sex, user_birthday, user_mobile from users;
select * from users;
```
2. 查询表的部分列

例如：从玩家表中查询玩家 QQ 和昵称。
```
select user_qq, user_name from users;
```
3. 别名的使用

例如：从玩家表中查询玩家 QQ 和昵称，并显示为"玩家 QQ"和"玩家昵称"。
```
select user_qq as '玩家QQ', user_name as '玩家昵称' from users;
select user_qq '玩家QQ', user_name '玩家昵称' from users;
```
4. DISTINCT 关键字

作用：消除结果集中的重复行。

例如：显示参与游戏的玩家 QQ，要求参与多个游戏的玩家不重复显示 QQ。
```
select distinct user_qq from scores;
```
5. LIMIT 关键字

作用：指定结果集中数据的显示范围。

例如：显示玩家表中第 3~5 条数据。
```
select * from users limit 2, 3;
select * from users limit 3 --- 只显示前三条数据
```

3.4.3 条件查询

1. 普通条件查询

语法：

SELECT COL_LIST FROM TABLE_NAME [WHERE CONDITION_EXPRESSION]

例如：查询 QQ 号为 12301 的玩家信息。
```
select * from users where user_qq=12301;
```
例如：查询分数大于 2 500 分的数据。
```
select * from scores where score>2500;
```
<> --- 不等于

>= --- 大于等于

<= --- 小于等于

例如：查询游戏编号为 1 且分数大于 4 000 分的分数信息。
```
select * from scores where gno=1 and score>4000;
```
逻辑运算符有三种：并且（and）、或者（or）和非（not）。

例如：查询游戏编号为 1 和 2 的分数信息。
```
select * from scores where gno=1 or gno=2;
```
2. 模糊查询

例如：查询分数在 2 500（含）~3 000（含）的分数。
```
select * from scores where score>=2500 and score<=3000;
```

```sql
select * from scores where score between 2500 and 3000;
```
例如：查询分数不在 2 500（含）~3 000（含）的分数信息。
```sql
select * from scores where score not between 2500 and 3000;
```
例如：查询 1987 年 1 月 1 日~1992 年 7 月 31 日出生的玩家。
```sql
select * from users where user_birthday between '1987-01-01' and '1992-0731';
```

通配符表示：			
'_'	一个字符	Branch like 'L_'	
%	任意长度	Route_Code Like 'AMS-%'	
[]	指定范围内	Airbusno Like 'AB0[1-5]'	
[^]	不在括号中	Airbusno Like 'AB0[^]'	

例如：查询所有姓孙的玩家信息。
```sql
select * from users where user_name like '孙%';
```
例如：查询所有非姓孙的玩家信息。
```sql
select * from users where user_name not like '孙%';
```

3. 查询空值的运算符

例如：查询生日为空的 NULL 的玩家信息。
```sql
select * from users where use_birthday is null;
```
例如：查询生日不为 NULL 的玩家信息。
```sql
select * from users where user_birthday is not null;
```

3.4.4 结果排序

1. 对指定列进行排序（排序依据、排序方式）

语法：
```sql
SELECT CLO_LIST FROM TABLE_NAME ORDER BY ORDER_BY_LIST [ASC/DESC]
```
例如：查询分数表中编号为 1 的所有分数信息，并按照分数升序排序。
```sql
select * from scores where gno=1 order by score asc;
```
例如：查询分数表中编号为 1 的所有分数信息，并按照分数降序排序。
```sql
select * from score where gno=1 order by score desc;
```
2. 对多列进行排序（排序依据、排序方式、优先级）

例如：查询分数表中的所有信息，并按照游戏编号的升序和分数的降序进行排序。
```sql
select * from scores order by gno asc, score desc;
```

3.5 在 MySQL 数据库中删除数据

1. 删除单个表中的数据

使用 DELETE 语句从单个表中删除数据，语法格式为：
```sql
DELETE FROM <表名> [WHERE 子句] [ORDER BY 子句] [LIMIT 子句]
```
语法说明如下：

(1) <表名>：指定要删除数据的表名。
(2) ORDER BY 子句：可选项。表示删除时，表中各行将按照子句中指定的顺序进行删除。
(3) WHERE 子句：可选项。表示为删除操作限定删除条件，若省略该子句，则代表删除该表中的所有行。
(4) LIMIT 子句：可选项。用于告知服务器在控制命令被返回到客户端前被删除行的最大值。
【注意】在不使用 WHERE 条件时，将删除所有数据。

2. 删除表中的全部数据

【实例1】删除 tb_courses_new 表中的全部数据，输入的 SQL 语句和执行结果如下所示。

```
mysql> DELETE FROM tb_courses_new;
Query OK, 3 rows affected ( 0.12 sec )
mysql> SELECT * FROM tb_courses_new;
Empty set ( 0.00 sec )
```

3. 根据条件删除表中的数据

【实例2】在 tb_courses_new 表中，删除 course_id 为 4 的记录，输入的 SQL 语句和执行结果如下所示。

```
mysql> DELETE FROM tb_courses
    -> WHERE course_id=4;
Query OK, 1 row affected ( 0.00 sec )
mysql> SELECT * FROM tb_courses;
+-----------+-------------+--------------+------------------+
| course_id | course_name | course_grade | course_info      |
+-----------+-------------+--------------+------------------+
|         1 | Network     |            3 | Computer Network |
|         2 | Database    |            3 | MySQL            |
|         3 | Java        |            4 | Java EE          |
+-----------+-------------+--------------+------------------+
3 rows in set ( 0.00 sec )
```

由运行结果可以看出，course_id 为 4 的记录已经被删除。

单元 4

使用 Python 对 OpenStack 进行二次开发

学习目标

◎ 了解软件项目的开发流程。
◎ 掌握使用 PowerDesigner 工具对数据库进行设计。
◎ 掌握使用 Axure 工具对项目进行原型设计,并能在以后的学习、工作过程中熟练使用 PowerDesignr 和 Axure 工具。

4.1 需求分析

4.1.1 软件项目开发流程

软件开发流程即软件设计思路和方法的一般过程,是一个逐步渐进的过程,将整个软件开发过程划分为顺序相接的四个阶段,每个阶段完成全部规定的任务后再进入下一个阶段,一个软件从开始到最后一共需要以下四个阶段。

1. 初始需求阶段
(1) 用户提出需求:确定项目开发的目标和范围,与其可行性。
(2) 分析需求规格:确定主要功能模块,确定开发周期和报价。
(3) 需求讨论规划:双方面谈,将软件需要实现的各个功能进行详细需求分析调整。

2. 原型设计阶段
(1) 原型详细设计:将需求分析转化成未来系统符合用户期望的原型设计。
(2) 开需求评审会:原型设计完成后,客户审核并确认具体设计,供应商开始编写实现。

3. 开发测试阶段
(1) 软件开发设计:对整个软件系统进行设计,如系统框架设计、数据库设计等,为系统开发一个健壮的结构并调整设计使其与实现环境相匹配。

（2）程序开发编码：在开发构建阶段，由供应商程序员根据详细设计及计划，将所有应用程序功能开发并集成为产品。

（3）软件测试阶段：测试要验证对象间的交互作用，验证软件中所有组件的正确集成，检验所有的需求已被正确实现，识别并确认缺陷在软件部署之前被提出并处理。

4. 运行维护阶段

（1）产品软件部署：部署的目的是成功生成版本并将软件分发给最终用户。

（2）正式验收交付：要确定软件、环境、用户是否可以开始系统的运作，交付阶段的重点是确保软件对最终用户是可用的。

（3）后期项目维护：软件产品发布后，根据需求变化或硬件环境的变化对应用程序进行修改。软件开发中一系列操作以满足客户的需求并以解决客户的问题为目的。

本单元只涉及软件开发第一阶段中需求分析的相关内容。软件项目开发流程如图 4-1 所示。

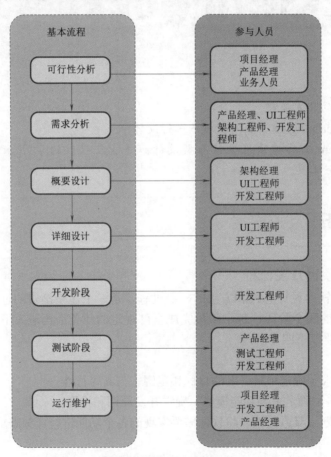

图 4-1 软件项目开发流程

4.1.2 项目需求分析

1. 基本概念

需求分析是指理解用户需求，就用户的功能需求与客户达成一致，并需要估计项目风险和评

单元 4　使用 Python 对 OpenStack 进行二次开发

估项目代价，最终形成开发计划的一个复杂过程。在这个过程中，用户是处于主导地位的，需求分析工程师和项目经理要负责整理用户需求，为之后的项目打下基础。

从广义上理解：需求分析包括需求的获取、分析、规格说明、变更、验证、管理等一系列需求工程。从狭义上理解：需求分析指需求的分析、定义过程。

需求分析阶段结束后应该得到相应的需求分析报告。

2. 分析内容

需要分析的内容可以包含：公司应用需求、技术资金投入与生产效益、行业技术发展趋势、国家政策支持等。

3. 分析过程

需求分析阶段的工作可以分为四方面：问题识别、分析与综合、指定规格说明、评审。

4. 分析方法

需求分析的方法有很多，如原型化方法、结构化方法和动态分析法等。

4.1.3　对 OpenStack 进行二次开发需求分析

1. 镜像管理

以列表的形式显示镜像相关信息，包括镜像名称、类型、状态、磁盘格式、大小、创建时间、修改时间、操作等信息，如图 4-2 所示。

图 4-2　镜像信息

（1）单击"创建镜像"按钮：进入创建镜像页面，如图 4-3 所示。

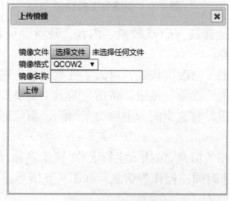

图 4-3　创建镜像

选择镜像文件以及镜像时，镜像名称自动生成，单击"上传"按钮，将镜像上传，并保存成功，刷新页面；其中镜像格式为下拉框，选项为 QCOW2、QCOW2C、VDI、VMDK、IMG、IOS。

(2) 在列表的操作栏中单击"修改"按钮，对镜像信息进行修改。

(3) 在列表的操作栏中单击"删除"按钮，将指定的镜像删除，删除后不可恢复。

(4) 单击"删除选中"按钮，将选中的镜像删除，删除后不可恢复。

2. 云主机类型管理

以列表的形式显示云主机类型相关信息，包括云主机类型名称、VCPU 数量、内存、硬盘大小、id、操作等信息，如图 4-4 所示。

图 4-4　云主机类型管理

(1) 单击"创建云主机类型"按钮：进入创建云主机类型页面，如图 4-5 所示。

图 4-5　创建云主机类型

输入云主机类型名称，并选择镜 VCPU 数量、内存、硬盘等信息后，单击"确定"按钮，保存云主机类型，并刷新列表页面。

(2) 在列表的操作栏中单击"修改"按钮，对云主机类型信息进行修改。

(3) 在列表的操作栏中单击"删除"按钮，将指定的云主机类型删除，删除后不可恢复。

(4) 单击"删除选中"按钮，将选中的云主机类型删除，删除后不可恢复。

3. 云主机运行管理

以列表的形式显示云主机相关信息，包括云主机名称、镜像名称、IP 地址、云主机状态、服务器、电源状态、所属配置状态、创建时间、操作等信息，如图 4-6 所示。

图 4-6　云主机运行管理

支持批量开启云主机操作、批量关闭云主机操作、批量重启云主机操作、批量挂起云主机操作、批量恢复云主机操作、批量删除云主机操作等。

（1）单击"开启选中"按钮，批量选中云主机进行开机。
（2）单击"关闭选中"按钮，批量选中云主机进行关机。
（3）单击"重启选中"按钮，批量选中云主机进行重启。
（4）单击"挂起选中"按钮，批量选中云主机进行挂起。
（5）单击"恢复选中"按钮，批量选中云主机进行恢复。
（6）单击"删除选中"按钮，批量选中云主机进行删除。

云主机列表中的操作为下拉菜单，选项有开机、关机、重启、挂起、恢复、初始化、远程等。

4.2　数据库设计

本章讲解对 OpenStack 进行二次开发，主要使用数据库为 OpenStack 中的三个数据库，分别是：glance、nova、nova_api，根据上面的需求内容，只对以上功能需要用到的数据库以及数据表进行讲解分析，其他数据库中的表在单元 8 综合案例中进行进一步讲解。

1. 数据库：glance

数据表：image。

作用：镜像表，存储 OpenStack 所有的镜像信息，如表 4-1 所示。

表 4-1　镜像信息

序号	字段名	类型	能否为空	主键	字段说明
1	id	varchar(36)	NO	PRI	主键
2	name	varchar(255)	YES		镜像名称
3	size	bigint(20)	YES		镜像大小
4	status	varchar(30)	NO		当前镜像状态
5	created_at	datetime	NO		创建时间
6	updated_at	datetime	YES		最后更新时间
7	deleted_at	datetime	YES		删除时间
8	deleted	tinyint	NO		是否已删除（1 为已删除，0 为未删除）
9	disk_format	varchar(20)	YES		镜像格式

2. 数据库：nova_api

数据表：flavors。

作用：云主机类型表，存储 OpenStack 云主机类型的信息，如表 4-2 所示。

表 4-2　云主机类型信息

序号	字段名	类型	能否为空	主键	字段说明
1	created_at	datetime	YES		创建时间
2	updated_at	datetime	YES		更新时间
3	name	varchar(255)	NO		云主机类型名称
4	id	int	NO	PRI	主键
5	memory_mb	int	NO		内存大小
6	vcpus	int	NO		CPU 个数
7	flavorid	varchar(255)	NO		云主机类型 id
8	root_gb	int	YES		磁盘大小

3. 数据库：nova

数据表：instances。

作用：云主机信息表，存储 OpenStack 所有云主机的信息，如表 4-3 所示。

表 4-3　云主机信息

序号	字段名	类型	能否为空	主键	字段说明
1	created_at	datetime	YES		创建时间
2	updated_at	datetime	YES		更新时间
3	deleted_at	datetime	YES		删除时间
4	id	int	NO	PRI	主键
5	image_ref	varchar(255)	YES		引用镜像 id
6	vm_state	varchar(255)	YES		当前状态
7	memory_mb	int	YES		内存大小
8	vcpus	int	YES		CPU 个数
9	hostname	varchar(255)	YES		云主机名称
10	host	varchar(255)	YES		所属结点
11	uuid	varchar(36)	NO		云主机 id
12	root_gb	int	YES		磁盘大小

4.3　技能训练

4.3.1　使用 PowerDesigner 进行数据库设计

PowerDesigner 是 Sybase 公司的 CASE 工具集，使用它可以方便地对管理信息系统进行分析设计，它几乎包括数据库模型设计的全过程。利用 PowerDesigner 可以制作数据流程图、概念数据

单元 4　使用 Python 对 OpenStack 进行二次开发

模型、物理数据模型,可以生成多种客户端开发工具的应用程序,还可为数据仓库制作结构模型,也能对团队设计模型进行控制。它可与许多流行的数据库设计软件,如与 PowerBuilder、Delphi 和 Visual Basic 等相配合使用,以缩短开发时间和使系统设计更优化。

1. 新建模型(见图 4-7)

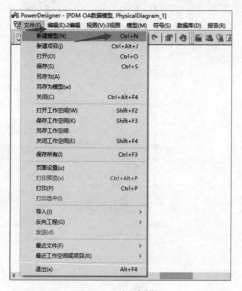

图 4-7　新建模型

2. 选择数据模型,填写模型名字,并选择使用的数据库(见图 4-8)

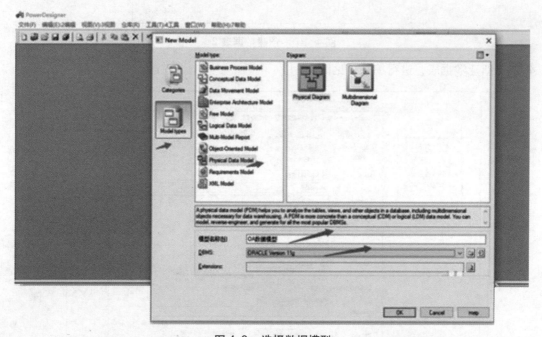

图 4-8　选择数据模型

3. 选择右侧工具栏，创建数据表（见图 4-9 和图 4-10）

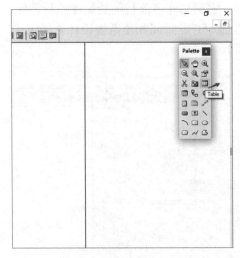

图 4-9　创建数据表 1

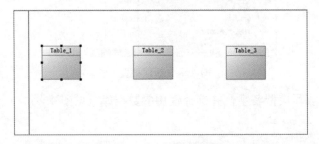

图 4-10　创建数据表 2

4. 双击新建数据表，进行编辑（见图 4-11）

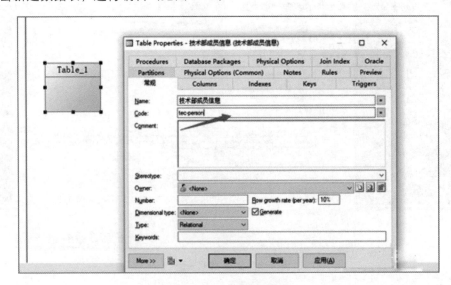

图 4-11　新建数据表

单元 4　使用 Python 对 OpenStack 进行二次开发

5. 建立索引（见图 4-12）

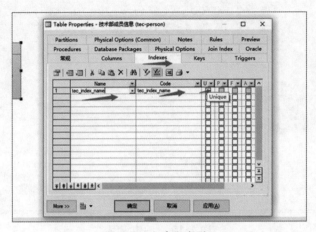

图 4-12　建立索引

6. 创建视图（见图 4-13）

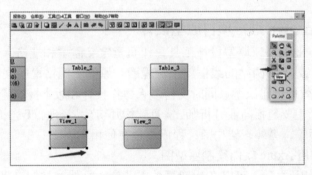

图 4-13　创建视图

7. 编辑视图（见图 4-14 和图 4-15）

双击视图进行编辑，在 SQL 选项窗口可以编辑视图的 SQL 语句，在 preview 中可以看到创建视图的完整语句。

图 4-14　编辑视图 1

图 4-15　编辑视图 2

同样，在左侧工具栏中可以创建所有数据。如果需要单个进行编辑直接单击某一项前面的图标即可。

4.3.2　使用 Axure 进行项目原型设计

Axure RP 是一款专业的快速原型设计工具，让负责定义需求和规格、设计功能和界面的专家能够快速创建应用软件或 Web 网站的线框图、流程图、原型和规格说明文档。作为专业的原型设计工具，它能快速、高效地创建原型，同时支持多人协作设计和版本控制管理。

Axure RP 的使用者主要包括商业分析师、信息架构师、产品经理、IT 咨询师、用户体验设计师、交互设计师、UI 设计师等，另外，架构师、程序员也在使用 Axure。

接下来介绍如何使用 Axure 软件绘制原型图。

（1）在打开 Axure RP 之后，可以在欢迎页单击"新建文件"按钮（见图 4-16），这样便可以开始建立一个新的项目（见图 4-17）。

图 4-16　新建文件

单元 4　使用 Python 对 OpenStack 进行二次开发

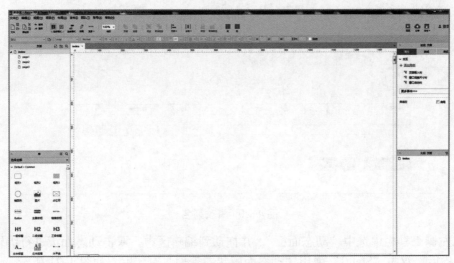

图 4-17　新建项目

（2）在页面上右击，在弹出的快捷菜单中选择"重命名"命令，如图 4-18 所示。

图 4-18　重命名

（3）在左侧工具栏中选中"表格"，并拖动到右侧编辑区中，默认为 3 行 3 列，在表格上右击，可以在指定位置插入行或者列，如图 4-19 所示。

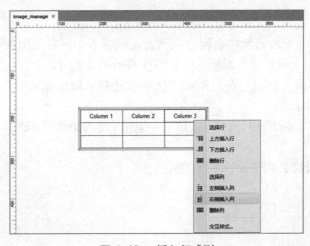

图 4-19　插入行或列

(4) 根据项目内容插入需要的列，并修改列名，如图 4-20 所示。

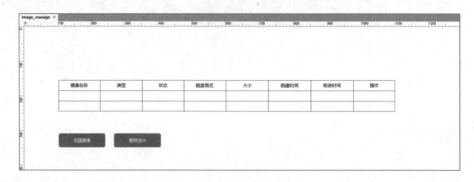

图 4-20 修改列名

(5) 从左侧工具栏中选中"动态面板"，并拖动到编辑区中，双击动态面板进行编辑，输入该动态面板名称后，双击"State1"弹出"动态面板状态管理"对话框，如图 4-21 所示。

(6) 在左侧工具栏中选择相应的工具，拖到右侧动态面板编辑区中，对页面内容进行排版编辑，如图 4-22 所示。

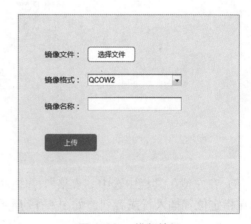

图 4-21 "动态面板状态管理"对话框　　　　图 4-22 排版编辑

(7) 编辑完成动态面板后回到主页面，在动态面板上右击，在弹出的快捷菜单中选择"设为隐藏"命令（见图 4-23），可使动态面板的内容不在初始化页面时显示。

(8) 选中"创建镜像"按钮，在右侧编辑区中，选择"鼠标单击时"，为创建镜像按钮添加单击事件，如图 4-24 所示。

(9) 在弹出页面的"Case1"中，选择"显示——编辑（动态面板）"，单击"确定"按钮保存设置，如图 4-25 所示。

(10) 使用快捷键【F5】预览设计完成的页面。

单元 4　使用 Python 对 OpenStack 进行二次开发

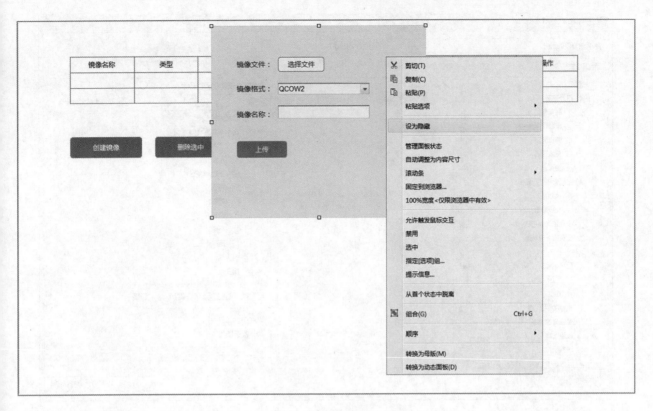

图 4-23　隐藏设置

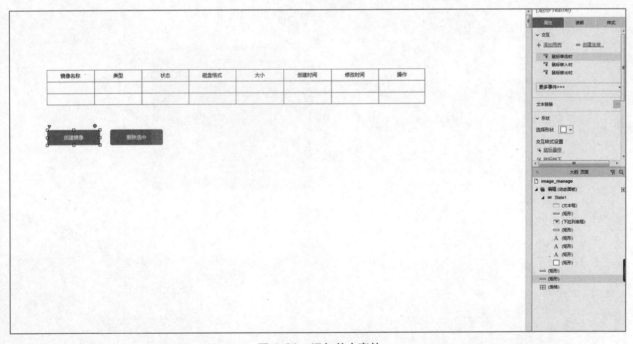

图 4-24　添加单击事件

图 4-25　保存设置

单元 5

OpenStack 基础配置

学习目标

◎ 掌握 OpenStack 环境搭建所需要的组件和服务，以及硬件条件；
◎ 掌握 OpenStack 架构，并能进行规划设计，特别需要进行完善的网络规划、数据库设置、时间同步等。

5.1 OpenStack 环境准备工作

从理论上，了解了什么是 OpenStack，了解到 OpenStack 能做些什么。但是只有原理，没有实践操作，对于大家来说还是只能在云里雾里看花，模糊不清，并不能真正看到云计算的实质内容。下面将动手安装、配置 OpenStack，并进行应用，完整地了解 OpenStack 是如何工作的。

在本书的单元 2 和单元 3 中介绍了 CentOS 操作系统，在 CentOS 操作系统中 MySQL 数据库的基本操作及配置方法，都是为学习安装及配置 OpenStack 做好准备。

5.1.1 OpenStack 实验部署架构

此次用来部署的 OpenStack（Rocky 版本）是一个开源的云计算平台，仅用作部署实验环境，其部署架构如图 5-1 所示，目的是为了让大家掌握主要的 OpenStack 服务及组件。如果需要在生产环境中部署，需要考虑更加复杂的生产环境体系。

5.1.2 OpenStack 实验环境硬件需求

运行 OpenStack 实验环境，需要至少两个核心结点（一个为控制结点，另一个为计算结点），这里所说的结点可以理解为两台物理服务器主机。当然在部署实验环境时可能没有服务器，也可以使用两台虚拟机来模拟物理服务器主机进行 OpenStack 实验环境的部署，其硬件需求如图 5-2 所示。

云计算技术与应用

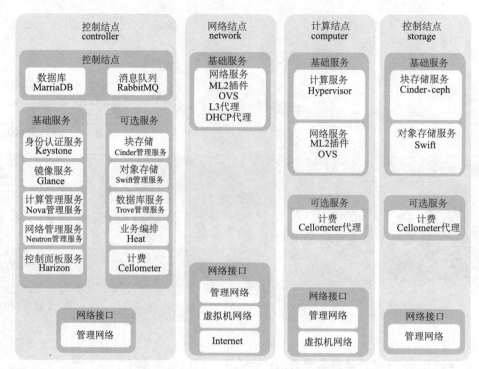

图 5-1 OpenStack 实验环境部署架构

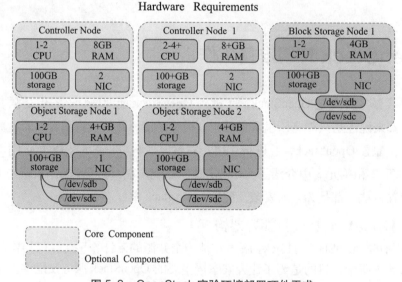

图 5-2 OpenStack 实验环境部署硬件需求

图 5-2 中核心组件为实心方框，可选组件为虚线方框。

1. 核心组件

（1）控制结点（Controller Node）：控制结点需要负责对其他结点的控制，如虚拟机的建立、网络的分配、存储的分配等。运行身份认证服务、镜像服务。还包含一些支撑服务，如数据库、消息队列和 NTP。控制结点至少需要两个网络结点。根据图 5-2 所示的 OpenStack 实验环境部署硬件需求，最基本的配置环境需要 8 GB 内存、100 GB 存储、2 个网卡。

（2）计算结点（Compute Node）：计算结点主要运行虚拟机、网络服务代理。根据图 5-2 所示的 OpenStack 实验环境部署硬件需求，最基本的配置环境需要 8 GB 以上内存、100 GB 以上存储、2 个网卡。

2. 可选组件

（1）块存储结点：块存储结点主要提供的是块存储。什么是块存储呢？例如通过虚拟出一块磁盘，挂载到虚拟机上，这个操作就像是给虚拟机新加了一块硬盘，同样可以对这个磁盘进行挂载、卸载、格式化、文件转换等操作。当虚拟机空间不足时，可通过这种方法进行扩展。

（2）对象存储结点：对象存储服务主要提供的是对象存储。什么是对象存储呢？例如可以虚拟出一部分磁盘空间，这个空间只能存放文件，不能进行格式化、文件转换。

5.1.3　修改结点名称

在配置路由器和交换机时，需要将路由器和交换机的名称定义好，再进行配置，养成一个良好的配置习惯非常重要。

重命名控制结点的主机名为 controller（该命令只适用于 CentOS7）：

```
[root@wz~]#hostnamectl set-hostname controller      //新的主机名称
                                                    //使用这个命令会立即生效且重启也生效
```

重命名计算结点的主机名为 compute：

```
[root@wz~]#hostnamectl set-hostname compute
```

如果在机器重启后主机名不对，或者后面在配置时报错。请检查主机名是否正确，若有差异，使用下面的命令进行修改：

```
vim /etc/hosts
```

加入如下配置：

```
10.0.0.9 controller
10.0.0.8 compute
vim /etc/hostname
```

删除原来的信息加入以下配置：

```
controller
vim /etc/sysconfig/network
```

加入以下配置：

```
HOSTNAME=controller
```

计算结点的配置同上。

5.1.4　安全设置

为了避免 OpenStack 安装配置过程中出现问题，导致某些步骤失败，这里将 CentOS 操作系统的防火墙和 SELinux 安全服务关闭。命令如下（两个结点均需运行）：

```
[root@wz~]#systemctl stop firewalld.service         //停止 firewall
[root@wz~]#systemctl disable firewalld.service      //禁止 firewall 开机启动
```

【注意】CentOS7 的防火墙改用 firewalld。

[root@wz~]#setenforce 0   //临时关闭

[root@wz~]#sed -i 's/=enforcing/=disabled/' /etc/selinux/config // /永久关闭 [root@wz~]#reboot   //重启系统命令

5.1.5 配置 Yum 源

官网中的 Yum 源安装软件包时用的是国外的源，下载速度较慢，如果软件包非常大则非常耗时。国内阿里云提供相应的安装源，可以配置阿里云的 Yum 源来提高下载安装速度。配置如下（两个结点均需配置）：

1. 将默认 Yum 源备份

[root@controller~]# mkdir /opt/centos-yum.bak

[root@controller~]# mv /etc/yum.repos.d/* /opt/centos-yum.bak/

2. 查看系统版本

[root@controller~]# cat /etc/redhat-release   //查看系统的版本

3. 下载阿里云 Yum 源 repo 文件（对应自己的系统版本下载即可）

CentOS 7 中的命令如下：

#wget -O /etc/yum.repos.d/CentOS-Base.repo http://mirrors.aliyun.com/repo/Centos-7.repo

5.2 主机网络设置

OpenStack 中结点之间都需要网络通信的支持。如果没有网络，控制结点、计算结点和存储结点之间无法交互，OpenStack 可以说就是瘫痪的。根据图 5-3 所示的部署架构，部署在每个结点上的操作系统，必须配置网络接口。网络接口配置完成后，一定要保证结点之间的网络能够互相通信，并处于同一网络中。另外，由于管理要求，如进行程序包的安装、安全更新、DNS、NTP 等操作，结点也都需要 Internet 访问权限。

（1）管理网络（内网）的网段地址及网关信息如下：

网关：192.168.0.1。

控制结点 IP：10.0.0.9（网卡 1）。

计算结点 IP：192.168.0.62（网卡 1）。

管理网络的网段必须是在一个子网内，并保证能互相通信。管理网络还需要一个网关地址为所有结点提供 Internet 访问，用于结点管理目的，例如软件包安装，安全更新，DNS 和 NTP 服务。

（2）专用网络（外网，可连接 Internet）的网段地址：

网关：192.168.1.1。

控制结点 IP：192.168.1.61（网卡 2）。

计算结点 IP：192.168.1.62（网卡 2）。

专用网络地址段 192.168.1.200~192.168.1.210 为 OpenStack 环境中的实例提供 Internet 访问。

单元 5　OpenStack 基础配置

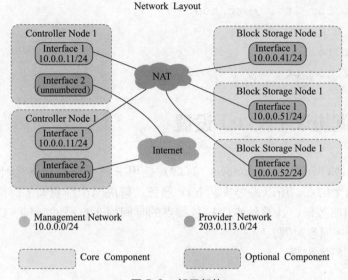

图 5-3　部署架构

1. 控制结点（Controller）网络配置

网卡 1 设置：

[root@controller~]#vi /etc/sysconfig/network-scripts/ifcfg-网卡 1 名称
TYPE=Ethernet
PROXY_METHOD=none
BROWSER_ONLY=no
BOOTPROTO=static　　#IP 获取方式为静态 IP
DEFROUTE=yes
IPV4_FAILURE_FATAL=no
IPV6INIT=no
IPV6_AUTOCONF=yes
IPV6_DEFROUTE=yes
IPV6_FAILURE_FATAL=no
IPV6_ADDR_GEN_MODE=stable-privacy
NAME=eno1
DEVICE=eno1
ONBOOT=yes　　#激活网卡
IPADDR=10.0.0.9　　#IP 地址 1
PREFIX=24
GATEWAY=192.168.0.1 #IP 网关
DNS1=114.114.114.114　　#IP DNS

2. 计算结点（Compute）网络配置

网卡 2 设置（请参考网卡 1 进行设置）：

还需要测试网络的连通性，并添加主机名映射（两个结点）：

```
[root@controller~]#vi /etc/hosts
```
在末尾添加命令如下：
```
10.0.0.9 controller
192.168.0.62 compute
```

5.3 网络时间协议（NTP）设置

网络时间协议（Network Time Protocol，NTP）是用来同步网络中各个计算机时间的协议。由于本次安装涉及两个结点，所以需要安装NTP服务，解决两个结点间的时间同步问题。其中，以控制结点作为时间服务器，计算结点以控制结点的时间同步自己的时钟。CentOS 7.2开始改用chrony命令行实用程序同步时间。

1. 在控制结点进行的配置

安装软件包：
```
[root@controller~]#yum install chrony -y
```
修改配置文件，使其时间与阿里云的时间同步：
```
[root@controller~]#vi /etc/chrony.conf
server ntp1.aliyun.com iburst
allow 10.0.0/24
server 10.0.0.9 iburst
```
重启服务，并配置开机启动：
```
[root@controller ~]# systemctl restart chronyd
[root@controller ~]# systemctl enable chronyd
```
设置时区，同步时间：
```
timedatectl set-timezone Asia/Shanghai
chronyc sources
```
现在使用timedatectl命令检查是否启用了NTP同步并且它是否实际同步：
```
[root@compute ~]# timedatectl
```

2. 在计算结点进行的配置

安装软件包：
```
[root@compute ~]#yum install chrony -y
```
修改配置文件，使其与控制结点的时间同步：
```
[root@compute ~]#vi /etc/chrony.conf
server controller iburst
```
重启服务，并配置开机启动：
```
[root@compute ~]# systemctl restart chronyd
[root@compute ~]# systemctl enable chronyd
```
设置时区，同步时间：
```
timedatectl set-timezone Asia/Shanghai
```

```
chronyc sources
```
现在使用 timedatectl 命令检查是否启用了 NTP 同步并且它是否实际同步：
```
[root@compute ~]# timedatectl
```
输入命令查看时间同步结果：
```
[root@compute ~]# chronyc sources
```

5.4 OpenStack 包的安装

完成 NTP 同步时间后，需要在控制结点和计算结点分别启用 OpenStack 存储库。

安装 Rocky 版本存储库，启用 OpenStack 存储库。

在两个结点上进行配置，这里以 Controller 结点为例：

(1) 在 CentOS 上，只需安装 Rocky 版本软件包即可启用 OpenStack 存储库。
```
[root@controller~]# yum install centos-release-openstack-rocky -y
[root@controller~]#yum clean all
[root@controller~]# yum makecache
```
(2) 升级所有结点上的包。
```
[root@controller~]# yum upgrade -y
```
(3) 安装 OpenStack 客户端。
```
[root@controller~]#yum install python-openstackclient -y
```
(4) CentOS 默认启用 SELinux。安装 OpenStack-SELinux 软件包以自动管理 OpenStack 服务的安全策略。
```
[root@controller~]# yum install openstack-selinux -y
```

5.5 安装及设置 SQL 数据库

大多数 OpenStack 服务都是使用 SQL 数据库来存储信息，此次需要使用的数据库是 MariaDB 和 MySQL。OpenStack 服务还支持其他 SQL 数据库，包括 PostgreSQL。以下需要在控制结点上进行 MariaDB 和 MySQL 数据库的安装和配置（CentOS 7 默认数据库为 MariaDB）。

(1) 安装 MariaDB 和 MySQL 数据库。
```
[root@controller~]# yum install mariadb mariadb-server MySQL-python python2-PyMySQL -y
```
(2) 创建和编辑 /etc/my.cnf.d/openstack.cnf 文件（如果需要 /etc/my.cnf.d/ 文件，先将其备份）并完成以下操作：
```
vi /etcmy.cnf.d/mariadb_openstack.cnf
```
在 [mysqld] 配置段落，添加以下配置：
```
[mysqld]
bind-address=10.0.0.9
```

```
default-storage-engine=innodb
innodb_file_per_table=on
max_connections=4096
collation-server=utf8_general_ci
character-set-server=utf8
```
【注意】以上参数配置能保证 OpenStack 程序和 MySQL 数据库的正常运行。

(3) 完成安装。

启动数据库服务并将其配置为在系统引导时启动：

```
[root@controller~]# systemctl enable mariadb.service
[root@controller~]# systemctl start mariadb.service
```

(4) 运行 mysql_secure_installation 脚本初始化数据库服务，为数据库 root 账户设置密码，默认密码为空，然后输入密码 123456，一直按【y】键，并按【Enter】键确认。

```
[root@controller ~]# mysql_secure_installation
```

5.6 消息服务器设置

OpenStack 使用消息队列来协调服务之间的操作和状态信息。OpenStack 支持多种消息队列服务，包括 RabbitMQ、Qpid 和 ZeroMQ。因为大多数 OpenStack 发行版都支持 RabbitMQ，所以此次使用 RabbitMQ 消息队列服务，并进行安装和设置。消息队列服务通常在控制结点上运行。

(1) 安装软件包。

```
[root@controller~]# yum install rabbitmq-server -y
```

(2) 启动消息队列服务并设置服务开机启动。

```
[root@controller~]# systemctl enable rabbitmq-server.service
[root@controller~]# systemctl start rabbitmq-server.service
```

(3) 添加 OpenStack 用户，并设置密码。

```
[root@controller~]#rabbitmqctl add_user openstack openstack
```

(4) 为 OpenStack 用户增加配置、读取及写入相关权限。

```
[root@controller~]# rabbitmqctl set_permissions openstack ".*" ".*" ".*"
[root@controller~]#rabbitmqctl set_permissions -p "/" openstack ".*" ".*" ".*"
```

返回：Setting permissions for user "openstack" in vhost "/"。

5.7 安装及设置 Memcached

身份认证服务使用 Memcached 缓存令牌，Memcached 是一个高性能的分布式内存对象缓存系统。直接使用内存可行吗？如果直接使用内存，无法进行分布式扩展。为什么要用 Memcached 呢？举例说明，使用用于动态 Web 应用以减轻数据库负载，它通过在内存中缓存数据和对象来减少读取数据库的次数，从而提高动态、数据库驱动网站的速度。Memcached 服务通常在控制器结点上运行。

（1）安装软件包。

```
[root@controller~]# yum install memcached python-memcached -y
```

（2）编辑 /etc/sysconfig/memcached 文件并完成以下操作。

编辑 /etc/sysconfig/memcached，使用控制器结点来管理 IP 地址。通过管理网络启用其他结点的访问：

```
PORT="11211"
USER="memcached"
MAXCONN="8192"
CACHESIZE="3166"
OPTIONS="-l 127.0.0.1,controller"
```

其中，PORT 为端口号，MAXCONN 为最大连接数，CACHESIZE 为缓存数据大小，单位为兆 (M)，OPTIONS 为可选参数配置，如 -l 127.0.0.1，-1 代表数据不过期，永不失效（当然服务器关闭除外，因为 memcached 是缓存数据）。请根据实际情况进行修改。

（3）启动 Memcached 服务并将其配置为在系统引导时启动，完成安装。

```
[root@controller~]# systemctl enable memcached.service
[root@controller~]# systemctl start memcached.service
```

5.8 配置 Etcd

OpenStack 服务可以使用 Etcd，Etcd 比较多的应用场景是用于服务发现，服务发现（Service Discovery）要解决的是分布式系统中最常见的问题之一，即在同一个分布式集群中的进程或服务如何才能找到对方并建立连接。从本质上说，服务发现就是要了解集群中是否有进程在监听 upd 或者 tcp 端口，并且通过名字就可以进行查找和连接。Etcd 服务通常在控制器结点上运行。

（1）安装软件包。

```
[root@controller~]# yum install etcd -y
```

（2）编辑 /etc/etcd/etcd.conf 文件，以控制结点管理 IP 地址设置相关选项，并使其他结点通过管理网络进行访问。

```
[root@controller~]# vi /etc/etcd/etcd.conf
#[Member]
ETCD_DATA_DIR="/var/lib/etcd/default.etcd"
ETCD_LISTEN_PEER_URLS="http://10.0.0.9:2380"
ETCD_LISTEN_CLIENT_URLS="http://10.0.0.9:2379"
ETCD_NAME="controller"
```

```
#[Clustering]
ETCD_INITIAL_ADVERTISE_PEER_URLS="http://10.0.0.9:2380"
ETCD_ADVERTISE_CLIENT_URLS="http://10.0.0.9:2379"
ETCD_INITIAL_CLUSTER="controller=http://10.0.0.9:2380"
ETCD_INITIAL_CLUSTER_TOKEN="etcd-cluster-01"
ETCD_INITIAL_CLUSTER_STATE="new"
```

(3) 启动 Etcd 服务并设为开机启动。

```
[root@controller~]# systemctl enable etcd
[root@controller~]# systemctl start etcd
```

单元 6

安装 OpenStack 服务

学习目标

◎ 掌握 OpenStack 环境搭建所需要的组件和服务；
◎ 掌握身份认证服务 Keystone、镜像服务 Glance、计算服务 Nova、网络服务 Neutron、Dashboard、块存储服务 Cinder 等内容的架构，以及安装配置

6.1 身份认证服务 Keystone 的安装配置

部署身份认证服务，依托 OpenStack 核心认证模块（Keystone），这里需要了解什么是 Keystone，其认证是怎么服务企业平台，包含的关键概念点，其中角色、租户和用户之间的职责分配；根据项目要求，需要为企业创建三个项目部门组、100 个用户；完成企业用户、租户的创建，同时绑定相对应的角色和租户权限。图 6-1 所示为 Keystone 用户管理。

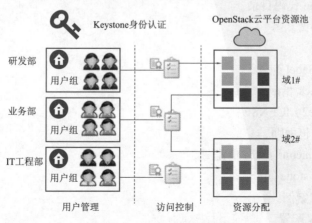

图 6-1　Keystone 用户管理

Keystone 组件架构如图 6-2 所示。

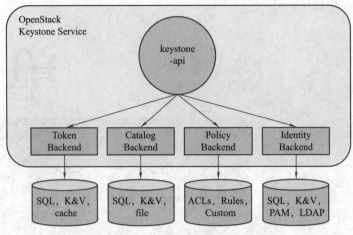

图 6-2　Keystone 组件架构

其特点如下：

Keystone 提供统一的、完整的 OpenStack 身份验证、服务目录、令牌、访问策略服务。

Keystone 处理所有的 API 请求，提供可配置的身份验证、服务目录、令牌、访问策略服务。

Keystone 标准的后端包括 SQL、LDAP、K&V 等。

大多数用户会使用定制化的 Keystone 作为 OpenStack 的认证服务。

6.1.1　Keystone 目录结构

Keystone 目录结构如下：

Bin：Keystone 服务执行文件。

Doc：技术文档。

Etc：Keystone 配置文件样例。

Examples：Keystone 使用样例。

Httpd：Keystone 使用 Apache 配置。

Keystone：Keystone 代码目录。

Tests：Keystone 测试。

Tools：工具。

Babel.cfg：Flask-Babel 配置。

HACKING.rst：Hack 指南。

LICENSE：Apache2 LICENSE。

MANIFEST.in：打包规则。

OpenStack-common.conf：OSLO 配置。

README.rst：Keystone 简介。

Run_tests.sh：测试案例。

Setup.cfg：setup.py 配置。

Setup.py：Keystone 安装脚本。

TODO：代办列表。

Tox.ini：Python 的标准化测试。

6.1.2 安装和配置组件

安装和配置认证服务只在 Controller 控制结点中进行。

1. 安装认证服务

运行以下命令安装 Keystone 和 Python-KeystoneClient，如图 6-3 和图 6-4 所示。

```
# yum install -y openstack-keystone python-keystoneclient openstack-utils
# yum install -y openstack-keystone httpd mod_wsgi
```

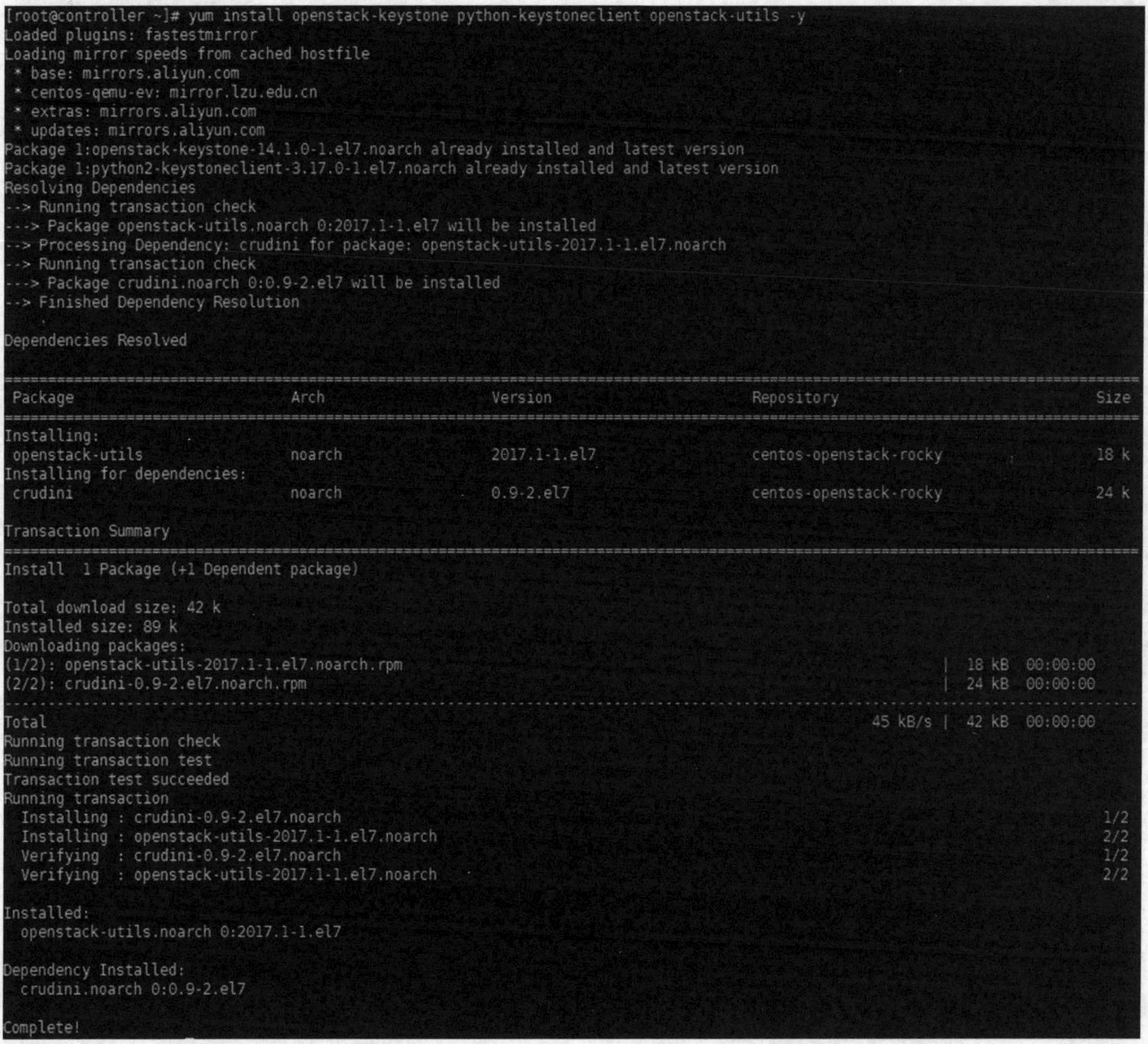

```
[root@controller ~]# yum install openstack-keystone python-keystoneclient openstack-utils -y
Loaded plugins: fastestmirror
Loading mirror speeds from cached hostfile
 * base: mirrors.aliyun.com
 * centos-qemu-ev: mirror.lzu.edu.cn
 * extras: mirrors.aliyun.com
 * updates: mirrors.aliyun.com
Package 1:openstack-keystone-14.1.0-1.el7.noarch already installed and latest version
Package 1:python2-keystoneclient-3.17.0-1.el7.noarch already installed and latest version
Resolving Dependencies
--> Running transaction check
---> Package openstack-utils.noarch 0:2017.1-1.el7 will be installed
--> Processing Dependency: crudini for package: openstack-utils-2017.1-1.el7.noarch
--> Running transaction check
---> Package crudini.noarch 0:0.9-2.el7 will be installed
--> Finished Dependency Resolution

Dependencies Resolved

================================================================================
 Package              Arch          Version            Repository              Size
================================================================================
Installing:
 openstack-utils      noarch        2017.1-1.el7       centos-openstack-rocky  18 k
Installing for dependencies:
 crudini              noarch        0.9-2.el7          centos-openstack-rocky  24 k

Transaction Summary
================================================================================
Install  1 Package (+1 Dependent package)

Total download size: 42 k
Installed size: 89 k
Downloading packages:
(1/2): openstack-utils-2017.1-1.el7.noarch.rpm           |  18 kB  00:00:00
(2/2): crudini-0.9-2.el7.noarch.rpm                      |  24 kB  00:00:00
--------------------------------------------------------------------------------
Total                                            45 kB/s |  42 kB  00:00:00
Running transaction check
Running transaction test
Transaction test succeeded
Running transaction
  Installing : crudini-0.9-2.el7.noarch                                     1/2
  Installing : openstack-utils-2017.1-1.el7.noarch                          2/2
  Verifying  : crudini-0.9-2.el7.noarch                                     1/2
  Verifying  : openstack-utils-2017.1-1.el7.noarch                          2/2

Installed:
  openstack-utils.noarch 0:2017.1-1.el7

Dependency Installed:
  crudini.noarch 0:0.9-2.el7

Complete!
```

图 6-3　安装认证 1

```
[root@controller ~]# yum install openstack-keystone httpd mod_wsgi -y
Loaded plugins: fastestmirror
Loading mirror speeds from cached hostfile
 * base: mirrors.aliyun.com
 * centos-qemu-ev: mirror.lzu.edu.cn
 * extras: mirrors.aliyun.com
 * updates: mirrors.aliyun.com
Resolving Dependencies
--> Running transaction check
---> Package httpd.x86_64 0:2.4.6-89.el7.centos will be installed
--> Processing Dependency: httpd-tools = 2.4.6-89.el7.centos for package: httpd-2.4.6-89.el7.centos.x86_64
--> Processing Dependency: /etc/mime.types for package: httpd-2.4.6-89.el7.centos.x86_64
--> Processing Dependency: libaprutil-1.so.0()(64bit) for package: httpd-2.4.6-89.el7.centos.x86_64
--> Processing Dependency: libapr-1.so.0()(64bit) for package: httpd-2.4.6-89.el7.centos.x86_64
---> Package mod_wsgi.x86_64 0:3.4-18.el7 will be installed
---> Package openstack-keystone.noarch 1:14.1.0-1.el7 will be installed
--> Processing Dependency: python-keystone = 1:14.1.0-1.el7 for package: 1:openstack-keystone-14.1.0-1.el7.noarch
--> Running transaction check
---> Package apr.x86_64 0:1.4.8-3.el7_4.1 will be installed
```

图 6-4　安装认证 2

2. 配置数据库连接

认证服务使用数据库来存储信息，需要在 Keystone 的配置文件中指定数据库的位置。这里将使用控制结点上的 MySQL 数据库，数据库用户名为 keystone，密码为 000000。

#openstack-config --set /etc/keystone/keystone.conf database connection mysql://keystone:000000@controller/keystone

(1) 创建数据库用户 keystone，密码为 123456。命令执行结果如图 6-5 所示。

mysql -u root -p

Enter password:　　# 输入 MySQL 的 root 用户的密码 123456。

mysql> CREATE DATABASE keystone;

mysql> GRANT ALL PRIVILEGES ON keystone.* TO 'keystone'@'localhost' IDENTIFIED BY 'keystone';

mysql> GRANT ALL PRIVILEGES ON keystone.* TO 'keystone'@'%' IDENTIFIED BY 'keystone';

mysql> flush privileges; （需要执行 flush privileges 命令刷新）

mysql> exit

```
[root@controller ~]# mysql -p123456
Welcome to the MariaDB monitor.  Commands end with ; or \g.
Your MariaDB connection id is 3
Server version: 10.1.20-MariaDB MariaDB Server

Copyright (c) 2000, 2016, Oracle, MariaDB Corporation Ab and others.

Type 'help;' or '\h' for help. Type '\c' to clear the current input statement.

MariaDB [(none)]> CREATE DATABASE keystone;
Query OK, 1 row affected (0.00 sec)

MariaDB [(none)]> GRANT ALL PRIVILEGES ON keystone.* TO 'keystone'@'localhost' IDENTIFIED BY 'keystone';
Query OK, 0 rows affected (0.00 sec)

MariaDB [(none)]> GRANT ALL PRIVILEGES ON keystone.* TO 'keystone'@'%' IDENTIFIED BY 'keystone';
Query OK, 0 rows affected (0.00 sec)

MariaDB [(none)]> flush privileges;
Query OK, 0 rows affected (0.00 sec)
```

图 6-5　命令执行结果

(2) 快速修改 Keystone 配置。

下面使用的快速配置方法需要安装 OpenStack-utils 才可以实现。

```
# openstack-config --set /etc/keystone/keystone.conf database connection mysql+pymysql://keystone:keystone@controller/keystone
#openstack-config --set /etc/keystone/keystone.conf token provider fernet
# 注意：keystone 不需要连接 rabbitmq
# 查看生效的配置
grep '^[a-z]' /etc/keystone/keystone.conf
# keystone 不需要启动，通过 http 服务进行调用
```

3. 初始化同步 keystone 数据库

(1) 同步 Keystone 数据库（44 张）。

```
# su -s /bin/sh -c "keystone-manage db_sync" keystone
```

(2) 同步完成进行连接测试（见图 6-6）。

保证所有需要的表已经建立，否则后面可能无法进行下去。

```
# mysql -h10.0.0.9 -ukeystone -pkeystone -e "use keystone;show tables;"
```

图 6-6 连接测试

4. 初始化 Fernet 令牌库

以下命令无返回信息：

```
#keystone-manage fernet_setup --keystone-user keystone --keystone-group keystone
#keystone-manage credential_setup --keystone-user keystone --keystone-group keystone
```

5. 配置启动 Apache（httpd）

（1）修改 httpd 主配置文件（见图 6-7）。

```
#vim /etc/httpd/conf/httpd.conf +95
----------------------------------
ServerName controller
----------------------------------
```

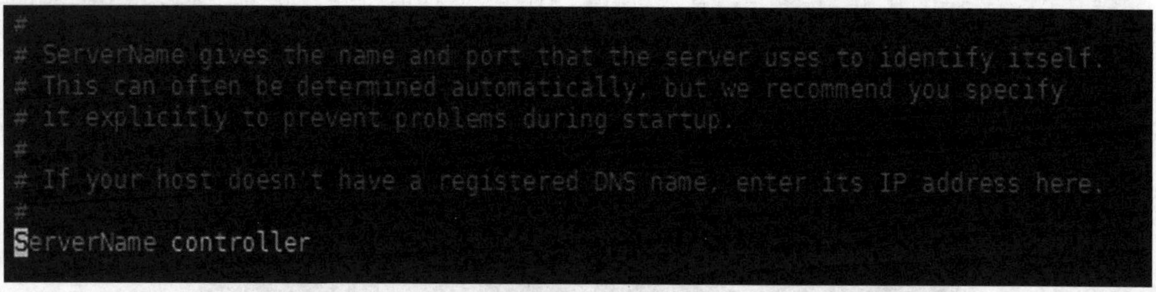

图 6-7 修改主配置文件

没有 vim 命令的可以安装 yum install-y vim。

（2）配置虚拟主机。

创建 Keystone 虚拟主机配置文件的快捷方式，也可以复制过来。

```
ln -s /usr/share/keystone/wsgi-keystone.conf /etc/httpd/conf.d/
```

（3）启动 httpd 并配置开机自启动（见图 6-8）。

```
#systemctl start httpd.service
#systemctl status httpd.service
#netstat -anptl|grep httpd
#systemctl enable httpd.service
#systemctl list-unit-files |grep httpd.service
```

如果 httpd 无法自启动，需要关闭 SELinux 或者安装 yum install OpenStack-SELinux。

6. 初始化 Keystone 认证服务

（1）创建 Keystone 用户、初始化的服务实体和 API 端点。

在（Queens 版本之前），引导服务需要两个端口提供服务（用户 5000 和管理 35357），R 本版本通过同一个端口提供服务。

```
[root@controller ~]# systemctl start httpd.service
[root@controller ~]# systemctl status httpd.service
● httpd.service - The Apache HTTP Server
   Loaded: loaded (/usr/lib/systemd/system/httpd.service; disabled; vendor preset: disabled)
   Active: active (running) since Thu 2019-05-09 14:01:48 CST; 23ms ago
     Docs: man:httpd(8)
           man:apachectl(8)
 Main PID: 16440 (httpd)
   Status: "Processing requests..."
   CGroup: /system.slice/httpd.service
           ├─16440 /usr/sbin/httpd -DFOREGROUND
           ├─16441 (wsgi:keystone- -DFOREGROUND
           ├─16442 (wsgi:keystone- -DFOREGROUND
           ├─16443 (wsgi:keystone- -DFOREGROUND
           ├─16444 (wsgi:keystone- -DFOREGROUND
           ├─16445 (wsgi:keystone- -DFOREGROUND
           ├─16446 /usr/sbin/httpd -DFOREGROUND
           ├─16447 /usr/sbin/httpd -DFOREGROUND
           ├─16448 /usr/sbin/httpd -DFOREGROUND
           ├─16449 /usr/sbin/httpd -DFOREGROUND
           └─16450 /usr/sbin/httpd -DFOREGROUND

May 09 14:01:48 controller systemd[1]: Starting The Apache HTTP Server...
May 09 14:01:48 controller systemd[1]: Started The Apache HTTP Server.
[root@controller ~]# netstat -anptl|grep httpd
tcp6       0      0 :::5000                 :::*                    LISTEN      16440/httpd
tcp6       0      0 :::80                   :::*                    LISTEN      16440/httpd
[root@controller ~]# systemctl enable httpd.service
Created symlink from /etc/systemd/system/multi-user.target.wants/httpd.service to /usr/lib/systemd/system/httpd.service.
[root@controller ~]# systemctl list-unit-files |grep httpd.service
httpd.service                                 enabled
```

图 6-8　启动 httpd

创建 Keystone 服务实体和身份认证服务，有三种类型，分别为公共的、内部的、管理的。

需要创建一个密码 ADMIN_PASS，作为登录 OpenStack 的管理员用户，这里创建密码为 123456。

以下为命令实例：

```
# keystone-manage bootstrap --bootstrap-password 123456 \
  --bootstrap-admin-url http://controller:5000/v3/ \
  --bootstrap-internal-url http://controller:5000/v3/ \
  --bootstrap-public-url http://controller:5000/v3/ \
  --bootstrap-region-id RegionOne
```

(2) 临时配置管理员账户的相关变量进行管理。

```
# export OS_PROJECT_DOMAIN_NAME=Default
# export OS_PROJECT_NAME=admin
# export OS_USER_DOMAIN_NAME=Default
# export OS_USERNAME=admin
# export OS_PASSWORD=123456
# export OS_AUTH_URL=http://controller:5000/v3
# export OS_IDENTITY_API_VERSION=3
```

查看声明的变量，如图 6-9 所示。

```
[root@controller ~]# env|grep OS_
```

```
[root@controller ~]# env|grep OS_
OS_USER_DOMAIN_NAME=Default
OS_PROJECT_NAME=admin
OS_IDENTITY_API_VERSION=3
OS_PASSWORD=123456
OS_AUTH_URL=http://controller:5000/v3
OS_USERNAME=admin
OS_PROJECT_DOMAIN_NAME=Default
```

图 6-9　查看声明变量

6.1.3　创建 Keystone 的一般实例

身份认证服务为每个 OpenStack 服务提供身份验证。身份认证服务使用域、项目、用户和角色的组合。

(1) 创建一个名为 example 的 Keystone 域。查看域如图 6-10 所示。

`# openstack domain create --description "An Example Domain" example`

```
[root@controller ~]# openstack domain create --description "An Example Domain" example
+-------------+----------------------------------+
| Field       | Value                            |
+-------------+----------------------------------+
| description | An Example Domain                |
| enabled     | True                             |
| id          | 8b94053e0ef64e1f8cf5ee233c395355 |
| name        | example                          |
| tags        | []                               |
+-------------+----------------------------------+
```

图 6-10　查看域

(2) 为 Keystone 系统环境创建名为 service 的项目提供服务，如图 6-11 所示。

`# openstack project create --domain default --description "Service Project" service`

```
[root@controller ~]# openstack project create --domain default --description "Service Project" service
+-------------+----------------------------------+
| Field       | Value                            |
+-------------+----------------------------------+
| description | Service Project                  |
| domain_id   | default                          |
| enabled     | True                             |
| id          | 0374894d62ee4870a322359e381f5e15 |
| is_domain   | False                            |
| name        | service                          |
| parent_id   | default                          |
| tags        | []                               |
+-------------+----------------------------------+
```

图 6-11　环境创建

(3) 创建 myproject 项目和对应的用户及角色，如图 6-12 所示。

`# openstack project create --domain default --description "Demo`

Project" myproject

```
[root@controller ~]# openstack project create --domain default --description "Demo Project" myproject
+-------------+----------------------------------+
| Field       | Value                            |
+-------------+----------------------------------+
| description | Demo Project                     |
| domain_id   | default                          |
| enabled     | True                             |
| id          | 03d6bbf7a0ff4d9da0567d3d523aefe5 |
| is_domain   | False                            |
| name        | myproject                        |
| parent_id   | default                          |
| tags        | []                               |
+-------------+----------------------------------+
```

图 6-12 创建 myproject 项目

(4) 在默认域创建 myuser 用户，如图 6-13 所示。

openstack user create --domain default --password-prompt myuser

这里的密码输入为 123456。

```
[root@controller ~]# openstack user create --domain default --password-prompt myuser
User Password:
Repeat User Password:
+---------------------+----------------------------------+
| Field               | Value                            |
+---------------------+----------------------------------+
| domain_id           | default                          |
| enabled             | True                             |
| id                  | ead92e22aba443dfa51895ced24b4723 |
| name                | myuser                           |
| options             | {}                               |
| password_expires_at | None                             |
+---------------------+----------------------------------+
```

图 6-13 创建 myuser 用户

(5) 在 role 表创建 myrole 角色，如图 6-14 所示。

openstack role create myrole

```
[root@controller ~]# openstack role create myrole
+-----------+----------------------------------+
| Field     | Value                            |
+-----------+----------------------------------+
| domain_id | None                             |
| id        | 1d9ad478281447acab0f68988f58b7c6 |
| name      | myrole                           |
+-----------+----------------------------------+
```

图 6-14 创建 myrole 角色

(6) 将 myrole 角色添加到 myproject 项目中和 myuser 用户组中，命令无返回信息。

openstack role add --project myproject --user myuser myrole

6.1.4 验证 Keystone 是否安装成功

（1）去除环境变量。关闭临时认证令牌机制，获取 token，验证 Keystone 配置是否成功。

```
# unset OS_AUTH_URL OS_PASSWORD
# env |grep OS_
```

（2）作为管理员用户去请求一个认证的 token。

测试是否可以使用 admin 账户进行登录认证，请求认证令牌，密码为 123456，如图 6-15 所示。

```
# openstack --os-auth-url http://controller:5000/v3 \
  --os-project-domain-name Default --os-user-domain-name Default \
  --os-project-name admin --os-username admin token issue
```

图 6-15　请求认证的 token

（3）使用普通用户获取认证 token。

```
# openstack --os-auth-url http://controller:5000/v3 \
  --os-project-domain-name Default --os-user-domain-name Default \
  --os-project-name myproject --os-username myuser token issue
```

6.1.5 创建 OpenStack 客户端环境脚本

前面使用了环境变量和命令选项的组合，通过 OpenStack 客户端与身份服务进行交互。为了提高客户端操作的效率，OpenStack 支持简单的客户端环境脚本，又称 OpenRC 文件。这些脚本通常包含所有客户机的通用选项，但也支持独特的选项。有关更多信息，请参见 OpenStack 用户指南。

1. 创建脚本

为 admin 和 myuser 项目和用户创建客户端环境脚本。本指南的未来部分引用这些脚本，为客户端操作加载适当的凭证。

客户端环境脚本的路径是不受限制的。为了方便起见，用户可以将脚本放置在任何位置，但要确保它们是可访问的。

（1）创建和编辑 keystone-admin-pass.sh 文件并添加以下内容：

```
export OS_PROJECT_DOMAIN_NAME=Default
export OS_USER_DOMAIN_NAME=Default
```

```
export OS_PROJECT_NAME=admin
export OS_USERNAME=admin
export OS_PASSWORD=123456
export OS_AUTH_URL=http://controller:5000/v3
export OS_IDENTITY_API_VERSION=3
export OS_IMAGE_API_VERSION=2
```

使用在身份认证服务中为 admin 账户选择的密码替换 ADMIN_PASS，这里改为 123456。

(2) 创建和编辑 demo-openrc 文件并添加以下内容：

```
export OS_PROJECT_DOMAIN_NAME=Default
export OS_USER_DOMAIN_NAME=Default
export OS_PROJECT_NAME=myproject
export OS_USERNAME=myuser
export OS_PASSWORD=myuser
export OS_AUTH_URL=http://controller:5000/v3
export OS_IDENTITY_API_VERSION=3
export OS_IMAGE_API_VERSION=2
```

使用在身份认证服务中为 myuser 用户选择的密码替换 MYUSER_PASS。

2. 使用脚本

作为一个特定的项目和用户运行客户端，可以在运行它们之前简单地加载相关的客户端环境脚本。

(1) 使用脚本加载相关客户端配置，以便快速使用特定租户和用户运行客户端。

```
source keystone-admin-pass.sh
```

(2) 请求身份验证令牌，如图 6-16 所示。

```
$ openstack token issue
```

```
[root@controller ~]# openstack token issue
+------------+---------------------------------------------------------------+
| Field      | Value                                                         |
+------------+---------------------------------------------------------------+
| expires    | 2019-05-09T07:44:32+0000                                      |
| id         | gAAAAABc08xQLfnsu3kcTMMhz4U8H6qXiOlF0G9FwttBELRwsylraM220t9eC-qtllvfEPL_JDBx38bnOZ-vDTFDydCTpDUNHRPL4lptBijMxKKoi_YQ
HG3sIhXUVOlrZGP4NTST9KbFcjSTquoe98CxPfx2uZ3-y41OvvIdqAzSbY9QSGnehRg |
| project_id | 066d7e4fcd9541db84b57199237a23a3                              |
| user_id    | 57d04499f59f4dcc8587e97fa1524a59                              |
+------------+---------------------------------------------------------------+
```

图 6-16 请求身份验证令牌

6.2 镜像服务 Glance 的安装配置

Glance 镜像服务实现发现、注册、获取虚拟机镜像和镜像元数据，镜像数据支持存储多种的存储系统，可以是简单文件系统、对象存储系统等。

Glance 镜像服务是典型的 C/S 架构，Glance 架构包括 Glance-Client、Glance、Glance Store。Glance 主要包括 REST API、数据库抽象层（DAL）、域控制器（Glance Domain Controller）、注册层（Registry Layer），Glance 使用集中数据库（Glance DB）在 Glance 各组件直接共享数据。所有的镜像文件操作通过 Glance-Store 库完成，Glance-Store 库提供了通用接口，对接后端外部不同存储。

Glance 组件架构如图 6-17 所示。

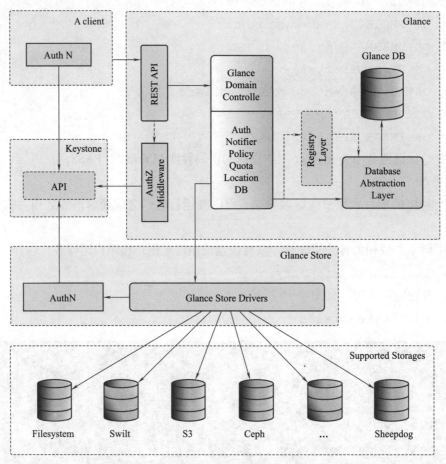

图 6-17 Glance 组件架构

其特点如下：

数据库中存放镜像文件的元数据。

存储镜像文件的实际后端可以有多种选择，可以使用 OpenStack 本身的组件 Swift、Cinder，也可以使用本地存储，或者使用 AWS 的 S3 等。

6.2.1 Glance 目录结构

Glance 目录结构如下：
Bin：Glance 服务执行文件。
Doc：技术文档。
Etc：Glance 配置文件样例。
Glance：Glance 代码目录。
Tools：工具。
Babel.cfg：Flask-Babel 配置。
HACKING.rst：Hack 指南。
LICENSE：Apache2 LICENSE。
MANIFEST.in：打包规则。
OpenStack-common.conf：OSLO 配置。
Pylintrc：Pylint 代码分析配置。
README.rst：Glance 简介。
Run_tests.sh：测试案例。
Setup.cfg：setup.py 配置。
Setup.py：Glance 安装脚本。
Tox.ini：Python 的标准化测试。

6.2.2 在控制端安装镜像服务 Glance

1. 安装镜像（见图 6-18）

创建数据库用户名 glance，密码为 glance。

```
# mysql -p123456
MariaDB [(none)]> CREATE DATABASE glance;
MariaDB [(none)]> GRANT ALL PRIVILEGES ON glance.* TO 'glance'@'localhost' IDENTIFIED BY 'glance';
MariaDB [(none)]> GRANT ALL PRIVILEGES ON glance.* TO 'glance'@'%' IDENTIFIED BY 'glance';
MariaDB [(none)]> flush privileges;
MariaDB [(none)]> exit
```

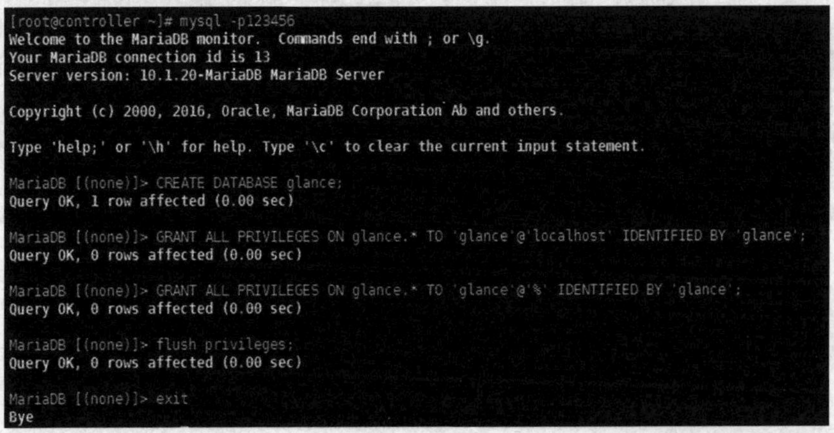

图 6-18 安装镜像

2. 在 Keystone 上注册 glance

(1) 在 Keystone 上创建 glance 用户，如图 6-19 所示。

```
# source keystone-admin-pass.sh
# openstack user create --domain default --password=glance glance
# openstack user list
```

```
[root@controller ~]# source keystone-admin-pass.sh
[root@controller ~]# openstack user create --domain default --password=glance glance
+---------------------+----------------------------------+
| Field               | Value                            |
+---------------------+----------------------------------+
| domain_id           | default                          |
| enabled             | True                             |
| id                  | a1a3a2cbd8e54cd298215daa65bd46d3 |
| name                | glance                           |
| options             | {}                               |
| password_expires_at | None                             |
+---------------------+----------------------------------+
[root@controller ~]# openstack user list
+----------------------------------+--------+
| ID                               | Name   |
+----------------------------------+--------+
| 57d04499f59f4dcc8587e97fa1524a59 | admin  |
| a1a3a2cbd8e54cd298215daa65bd46d3 | glance |
| ead92e22aba443dfa51895ced24b4723 | myuser |
+----------------------------------+--------+
```

图 6-19　创建 glance 用户

(2) 在 Keystone 上将 glance 用户添加为 service 项目的 admin 角色（权限）。

```
# openstack role add --project service --user glance admin
```

(3) 创建 glance 镜像服务的实体，如图 6-20 所示。

```
# openstack service create --name glance --description "OpenStack Image" image
# openstack service list
```

```
[root@controller ~]# openstack service create --name glance --description "OpenStack Image" image
+-------------+----------------------------------+
| Field       | Value                            |
+-------------+----------------------------------+
| description | OpenStack Image                  |
| enabled     | True                             |
| id          | d57d53c78ceb454c88aefcfb837f0ba6 |
| name        | glance                           |
| type        | image                            |
+-------------+----------------------------------+
[root@controller ~]# openstack service list
+----------------------------------+----------+----------+
| ID                               | Name     | Type     |
+----------------------------------+----------+----------+
| a2a28a2d9cc74c9399a67c3ed7372672 | keystone | identity |
| d57d53c78ceb454c88aefcfb837f0ba6 | glance   | image    |
+----------------------------------+----------+----------+
```

图 6-20　创建 glance 镜像服务实体

(4) 创建镜像服务的 API 端点（Endpoint），如图 6-21~ 图 6-24 所示。

```
# openstack endpoint create --region RegionOne image public http://10.0.0.9:9292
# openstack endpoint create --region RegionOne image internal http:
```

//10.0.0.9:9292
openstack endpoint create --region RegionOne image admin http://10.0.0.9:9292
openstack endpoint list

```
[root@controller ~]# openstack endpoint create --region RegionOne image public http://10.0.0.9:9292
+--------------+----------------------------------+
| Field        | Value                            |
+--------------+----------------------------------+
| enabled      | True                             |
| id           | 270a46d29e9548a1b6a8cc951072dc2c |
| interface    | public                           |
| region       | RegionOne                        |
| region_id    | RegionOne                        |
| service_id   | d57d53c78ceb454c88aefcfb837f0ba6 |
| service_name | glance                           |
| service_type | image                            |
| url          | http://10.0.0.9:9292             |
+--------------+----------------------------------+
```

图 6-21　创建镜像服务端点 1

```
[root@controller ~]# openstack endpoint create --region RegionOne image internal http://10.0.0.9:9292
+--------------+----------------------------------+
| Field        | Value                            |
+--------------+----------------------------------+
| enabled      | True                             |
| id           | f73c02998c394e74b7356dadf32f0ee7 |
| interface    | internal                         |
| region       | RegionOne                        |
| region_id    | RegionOne                        |
| service_id   | d57d53c78ceb454c88aefcfb837f0ba6 |
| service_name | glance                           |
| service_type | image                            |
| url          | http://10.0.0.9:9292             |
+--------------+----------------------------------+
```

图 6-22　创建镜像服务端点 2

```
[root@controller ~]# openstack endpoint create --region RegionOne image admin http://10.0.0.9:9292
+--------------+----------------------------------+
| Field        | Value                            |
+--------------+----------------------------------+
| enabled      | True                             |
| id           | 4f61a845d8944da4b860652a6e3a07bf |
| interface    | admin                            |
| region       | RegionOne                        |
| region_id    | RegionOne                        |
| service_id   | d57d53c78ceb454c88aefcfb837f0ba6 |
| service_name | glance                           |
| service_type | image                            |
| url          | http://10.0.0.9:9292             |
+--------------+----------------------------------+
```

图 6-23　创建镜像服务端点 3

```
[root@controller ~]# openstack endpoint list
+----------------------------------+-----------+--------------+--------------+---------+-----------+-----------------------------+
| ID                               | Region    | Service Name | Service Type | Enabled | Interface | URL                         |
+----------------------------------+-----------+--------------+--------------+---------+-----------+-----------------------------+
| 08a3168f6d454ad881d61fa1c1df8067 | RegionOne | keystone     | identity     | True    | internal  | http://controller:5000/v3/  |
| 270a46d29e9548a1b6a8cc951072dc2c | RegionOne | glance       | image        | True    | public    | http://10.0.0.9:9292        |
| 30b28bae2dcc4e2c876d8bd21137edc2 | RegionOne | keystone     | identity     | True    | admin     | http://controller:5000/v3/  |
| 4f61a845d8944da4b860652a6e3a07bf | RegionOne | glance       | image        | True    | admin     | http://10.0.0.9:9292        |
| b731ffd5ba8c493d8edfe4a35b6e2e33 | RegionOne | keystone     | identity     | True    | public    | http://controller:5000/v3/  |
| f73c02998c394e74b7356dadf32f0ee7 | RegionOne | glance       | image        | True    | internal  | http://10.0.0.9:9292        |
+----------------------------------+-----------+--------------+--------------+---------+-----------+-----------------------------+
```

图 6-24　创建镜像服务端点 4

3. 安装 Glance 相关软件

（1）安装 Glance 软件，如图 6-25 所示。

yum install openstack-glance python-glance python-glanceclient -y

```
[root@controller ~]# yum install openstack-glance python-glance python-glanceclient -y
Loaded plugins: fastestmirror
Loading mirror speeds from cached hostfile
 * base: mirrors.aliyun.com
 * centos-qemu-ev: mirror.lzu.edu.cn
 * extras: mirrors.aliyun.com
 * updates: mirrors.aliyun.com
Package 1:python2-glanceclient-2.13.1-1.el7.noarch already installed and latest version
Resolving Dependencies
--> Running transaction check
```

图 6-25　安装 glance 软件

（2）执行以下命令快速配置 glance-api.conf，如图 6-26 所示。

openstack-config --set /etc/glance/glance-api.conf database connection mysql+pymysql://glance:glance@controller/glance

openstack-config --set /etc/glance/glance-api.conf keystone_authtoken www_authenticate_uri http://controller:5000

openstack-config --set /etc/glance/glance-api.conf keystone_authtoken auth_url http://controller:5000

openstack-config --set /etc/glance/glance-api.conf keystone_authtoken memcached_servers controller:11211

openstack-config --set /etc/glance/glance-api.conf keystone_authtoken auth_type password

openstack-config --set /etc/glance/glance-api.conf keystone_authtoken project_domain_name Default

openstack-config --set /etc/glance/glance-api.conf keystone_authtoken user_domain_name Default

openstack-config --set /etc/glance/glance-api.conf keystone_authtoken project_name service

openstack-config --set /etc/glance/glance-api.conf keystone_authtoken username glance

openstack-config --set /etc/glance/glance-api.conf keystone_authtoken password glance

openstack-config --set /etc/glance/glance-api.conf paste_deploy flavor keystone

openstack-config --set /etc/glance/glance-api.conf glance_store

stores file,http
    # openstack-config --set /etc/glance/glance-api.conf glance_store default_store file
    # openstack-config --set /etc/glance/glance-api.conf glance_store filesystem_store_datadir /var/lib/glance/images/

图 6-26 快速配置 glance-api.conf

执行以下命令快速配置 glance-registry.con，如图 6-27 所示。
    # openstack-config --set /etc/glance/glance-registry.conf database connection mysql+pymysql://glance:glance@controller/glance
    # openstack-config --set /etc/glance/glance-registry.conf keystone_authtoken www_authenticate_uri http://controller:5000
    # openstack-config --set /etc/glance/glance-registry.conf keystone_authtoken auth_url http://controller:5000
    # openstack-config --set /etc/glance/glance-registry.conf keystone_authtoken memcached_servers controller:11211
    # openstack-config --set /etc/glance/glance-registry.conf keystone_authtoken auth_type password
    # openstack-config --set /etc/glance/glance-registry.conf keystone_authtoken project_domain_name Default
    # openstack-config --set /etc/glance/glance-registry.conf keystone_authtoken user_domain_name Default
    # openstack-config --set /etc/glance/glance-registry.conf keystone_authtoken project_name service
    # openstack-config --set /etc/glance/glance-registry.conf keystone_authtoken username glance
    # openstack-config --set /etc/glance/glance-registry.conf keystone_authtoken password glance
    # openstack-config --set /etc/glance/glance-registry.conf paste_

deploy flavor keystone

图 6-27 快速配置 glance-registry.con

查看生效的配置，如图 6-28 和图 6-29 所示。

[root@controller tools]# grep '^[a-z]' /etc/glance/glance-api.conf

图 6-28 查看生效配置 1

[root@controller tools]# grep '^[a-z]' /etc/glance/glance-registry.conf

图 6-29 查看生效配置 2

4. 同步 glance 数据库

（1）为 glance 镜像服务初始化同步数据库，如图 6-30 所示。

生成相关表：

su -s /bin/sh -c "glance-manage db_sync" glance

图 6-30　同步 glance 数据库

（2）同步完成后进行连接测试，如图 6-31 所示。
保证所有需要的表已经建立，否则后面可能无法进行下去。

```
# mysql -h10.0.0.9 -uglance -pglance -e "use glance;show tables;"
```

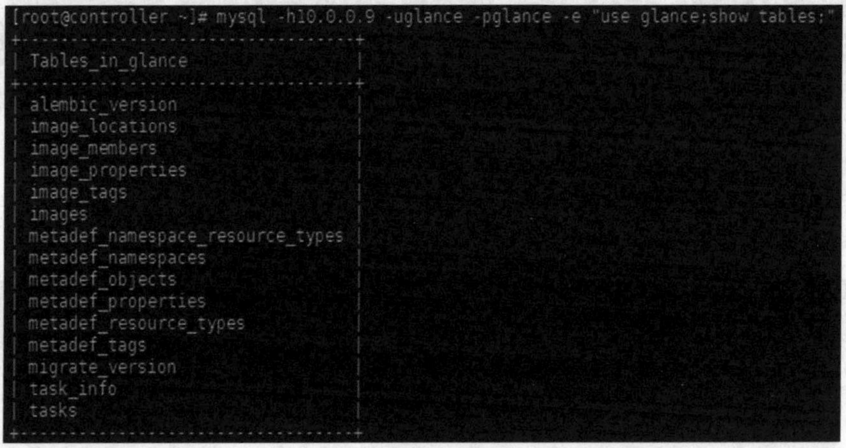

图 6-31　连接测试

5. 启动 glance 镜像服务

启动 glance 镜像服务，并配置开机自启动，如图 6-32~ 图 6-36 所示。

```
# systemctl start openstack-glance-api.service openstack-glance-
```

registry.service

```
[root@controller ~]# systemctl start openstack-glance-api.service openstack-glance-registry.service
```

图 6-32　启动镜像服务并配置 1

systemctl status openstack-glance-api.service openstack-glance-registry.service

```
● openstack-glance-api.service - OpenStack Image Service (code-named Glance) API server
   Loaded: loaded (/usr/lib/systemd/system/openstack-glance-api.service; disabled; vendor preset: disabled)
   Active: active (running) since Thu 2019-05-09 15:26:20 CST; 1s ago
 Main PID: 20432 (glance-api)
   CGroup: /system.slice/openstack-glance-api.service
           └─20432 /usr/bin/python2 /usr/bin/glance-api

May 09 15:26:21 controller glance-api[20432]: /usr/lib/python2.7/site-packages/paste/deploy/loadwsgi.py:22: DeprecationWar...ately.
May 09 15:26:21 controller glance-api[20432]: return pkg_resources.EntryPoint.parse("x=" + s).load(False)
May 09 15:26:21 controller glance-api[20432]: /usr/lib/python2.7/site-packages/paste/deploy/loadwsgi.py:22: DeprecationWar...ately.
May 09 15:26:21 controller glance-api[20432]: return pkg_resources.EntryPoint.parse("x=" + s).load(False)
May 09 15:26:21 controller glance-api[20432]: /usr/lib/python2.7/site-packages/paste/deploy/loadwsgi.py:22: DeprecationWar...ately.
May 09 15:26:21 controller glance-api[20432]: return pkg_resources.EntryPoint.parse("x=" + s).load(False)
May 09 15:26:21 controller glance-api[20432]: /usr/lib/python2.7/site-packages/paste/deploy/loadwsgi.py:22: DeprecationWar...ately.
May 09 15:26:21 controller glance-api[20432]: return pkg_resources.EntryPoint.parse("x=" + s).load(False)
May 09 15:26:21 controller glance-api[20432]: /usr/lib/python2.7/site-packages/paste/deploy/util.py:55: DeprecationWarning...filter
May 09 15:26:21 controller glance-api[20432]: val = callable(*args, **kw)
```

图 6-33　启动镜像服务并配置 2

```
● openstack-glance-registry.service - OpenStack Image Service (code-named Glance) Registry server
   Loaded: loaded (/usr/lib/systemd/system/openstack-glance-registry.service; disabled; vendor preset: disabled)
   Active: active (running) since Thu 2019-05-09 15:26:20 CST; 1s ago
 Main PID: 20433 (glance-registry)
   CGroup: /system.slice/openstack-glance-registry.service
           ├─20433 /usr/bin/python2 /usr/bin/glance-registry
           ├─20454 /usr/bin/python2 /usr/bin/glance-registry
           ├─20455 /usr/bin/python2 /usr/bin/glance-registry
           ├─20456 /usr/bin/python2 /usr/bin/glance-registry
           └─20457 /usr/bin/python2 /usr/bin/glance-registry

May 09 15:26:21 controller glance-registry[20433]: /usr/lib/python2.7/site-packages/paste/deploy/loadwsgi.py:22: Deprecatio...tely.
May 09 15:26:21 controller glance-registry[20433]: return pkg_resources.EntryPoint.parse("x=" + s).load(False)
May 09 15:26:21 controller glance-registry[20433]: /usr/lib/python2.7/site-packages/paste/deploy/loadwsgi.py:22: Deprecatio...tely.
May 09 15:26:21 controller glance-registry[20433]: return pkg_resources.EntryPoint.parse("x=" + s).load(False)
May 09 15:26:21 controller glance-registry[20433]: /usr/lib/python2.7/site-packages/paste/deploy/loadwsgi.py:22: Deprecatio...tely.
May 09 15:26:21 controller glance-registry[20433]: return pkg_resources.EntryPoint.parse("x=" + s).load(False)
May 09 15:26:21 controller glance-registry[20433]: /usr/lib/python2.7/site-packages/glance/registry/api/__init__.py:36: Dep...oval.
May 09 15:26:21 controller glance-registry[20433]: debtcollector.deprecate("Glance Registry service has been "
May 09 15:26:21 controller glance-registry[20433]: /usr/lib/python2.7/site-packages/paste/deploy/util.py:55: DeprecationWar...ilter
May 09 15:26:21 controller glance-registry[20433]: val = callable(*args, **kw)
Hint: Some lines were ellipsized, use -l to show in full.
```

图 6-34　启动镜像服务并配置 3

systemctl enable openstack-glance-api.service openstack-glance-registry.service

```
[root@controller ~]# systemctl enable openstack-glance-api.service openstack-glance-registry.service
Created symlink from /etc/systemd/system/multi-user.target.wants/openstack-glance-api.service to /usr/lib/systemd/system/openstack-glance-api.service.
Created symlink from /etc/systemd/system/multi-user.target.wants/openstack-glance-registry.service to /usr/lib/systemd/system/openstack-glance-registry.service.
```

图 6-35　启动镜像服务并配置 4

systemctl list-unit-files |grep openstack-glance*

```
[root@controller ~]# systemctl list-unit-files |grep openstack-glance*
openstack-glance-api.service              enabled
openstack-glance-registry.service         enabled
openstack-glance-scrubber.service         disabled
```

图 6-36　启动镜像服务并配置 5

6. 检查确认 glance 安装正确

(1) 下载镜像，如图 6-37 所示。

wget http://download.cirros-cloud.net/0.3.5/cirros-0.3.5-x86_64-disk.img

图 6-37 下载镜像

(2) 获取管理员权限。

source keystone-admin-pass.sh

(3) 上传镜像到 glance，如图 6-38 和图 6-39 所示。

openstack image create "cirros" --file cirros-0.3.5-x86_64-disk.img --disk-format qcow2 --container-format bare --public

图 6-38 上传镜像到 glance 1

图 6-39 上传镜像到 glance 2

(4) 查看镜像，如图 6-40 所示。

```
# openstack image list
```

图 6-40 查看镜像

6.3 计算服务 Nova 的安装配置

Nova 组件是最早的 OpenStack 组件之一，熟悉了 Nova 组件结构后，其他组件可以轻松学习。Nova 架构非常成熟，最早由 RackSpace 贡献，又称 RackSpace 架构。nova-api 支持 OpenStack 原生 API、AWSEC2 API，或者特殊的管理员 API（为特权用户执行管理操作）。它可强制执行一些默认的配额策略；可以通过 Keystone 认证，认证处理优先于内部任务处理；支持多种后端 Hypervisor 创建虚拟机，如 KVM、Xen、LXC、VMware、Hyper-V 等。

Nova，即计算服务，是 OpenStack 计算的弹性控制器。Nova 可以说是整个云平台最重要的组件，OpenStack 的其他组件依托 Nova，与 Nova 协同工作，组成整个 OpenStack 云平台。

Nova 由很多子服务组成，大家知道 OpenStack 是一个分布式系统，对于 Nova 这些服务会部署在两类结点上：计算结点和控制结点。计算结点上安装 Hypervisor，上面运行虚拟机，只有 nova-compute 需要放在计算结点上，其他子服务则是放在控制结点上的。

1. nova-api

nova-api 服务作为 Nova 组件对外的唯一窗口，接收和响应用户的 API 请求，当客户需要执行虚机相关的操作，能且只能向 nova-api 发送 REST API 请求，这里的客户包括终端用户、命令行和 OpenStack 其他组件。

Nova-api 服务支持 OpenStack Compute API、Amazon EC2 API，还有一个 Admin API 用于用户进行一些管理操作。

2. nova-api-metadata

Nova-api-metadata 主要用于接收虚拟机的元数据请求，通常在带有 nova-network 的多主机模式下才会使用。

3. nova-compute

Nova-compute 在计算结点上运行，它是一个工作进程（Worker Deamon），通过 Hypervisor APIs 来创建或者关闭虚拟机实例，这个过程相当复杂，总而言之，它从队列服务接收消息之后，启动虚拟机并在数据库中更新状态。

Nova-compute 的功能可以分为两类：一类是定时向 OpenStack 报告计算结点的状态；另一类是实现 instance 生命周期的管理。

4. nova-placement-api

主要用于追踪每个提供者的库存和使用量情况，如追踪计算结点的资源，存储池的使用情况以及 IP 的分配情况等。

5. nova-scheduler

如果有多个实体都能够完成任务,那么通常会有一个 scheduler 负责从这些实体中挑选一个最合适的来执行操作,nova-scheduler 就是处理队列中的请求,然后根据计算结点当时的资源使用情况选择一个最合适的计算结点来运行虚拟机,会通过消息中间件 rabbitMQ 向选定的计算结点发出 launch instance 的命令。

调度服务就好比是一个开发团队中的项目经理,当接到新的开发任务时,项目经理会评估任务的难度,考察团队成员目前的工作负荷和技能水平,然后将任务分配给最合适的开发人员。

6. nova-conductor

Nova-compute 需要获取和更新数据库中 instance 的信息,但 nova-compute 并不会直接访问数据库,而是通过 nova-conductor 实现数据的访问。在巨大的集群中,nova-conductor 可以水平扩展,但是不要部署在计算结点上。

如此设计的好处是让集群拥有更高的系统安全性和更好的系统伸缩性。

7. nova-cert

一个 Nova 的证书服务,提供 x509 证书支持,用于兼容 AWS。

8. nova-consoleauth

负责对访问虚拟机控制台请求提供 token 认证。

9. nova-novncproxy

提供基于 Web 浏览器的 VNC 访问。

10. nova-spicehtml5proxy

提供基于 HTML5 浏览器的 SPICE 访问。

11. nova-xvpvncproxy

提供基于 Java 客户端的 VNC 访问。

12. Message Queue

前文已提到 Nova 包含众多的子服务,这些子服务之间需要相互协调和通信,为解耦各个子服务,Nova 通过 Message Queue 作为子服务的信息中转站,因此在架构图上,子服务之间是没有直接的连线,它们都通过 Message Queue 联系。

13. SQL database

Nova 会存放一些云平台的状态数据在数据库中,如可用虚拟机类型、虚拟机是否使用中、网络的使用状态以及项目等。

这里通过一个 Nova 实例的启动过程来了解内部的数据和请求的运行情况。首先用户会通过 Web 界面或者 CLI 界面发送一个启动实例的请求,首先需要身份认证服务 Keystone 的请求,进行身份验证,通过之后将拿到的 token 向 nova-api 发送请求,验证 image 和 flavor 时存在。通过上述验证后将启动实例的请求发送给计算服务的调度结点,调度结点随机将此请求发送给一个计算结点让其启动实例,计算结点接受请求之后会到 Glance 存在的计算结点开始下载镜像并启动实例,计算结点启动虚拟机时会通过 Neutron 的 DHCP 获取一个对应的 IP 以及其他网络资源,再次在 OVS 网桥获取相应的端口绑定虚拟机的虚拟网卡接口。至此实例的启动已经完成。Nova 集合了多个服务进程,每个进程扮演着不同的角色。用户通过 REST API 接口访问,Nova 的内部服务通过 RPC 消息服务通信。API 服务产生数据读/写的 REST 请求,然后产生 REST 应答。内部服务通信通过

oslo.messaging 库 RPC 消息实现消息的通信。Nova 数据和请求的运行情况如图 6-41 所示。

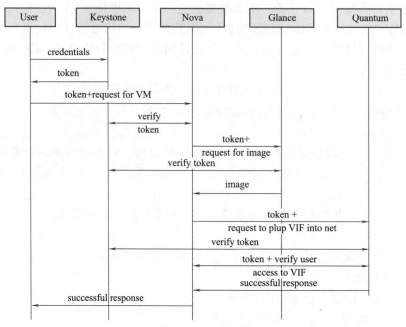

图 6-41 Nova 数据和请求的运行情况

OpenStack 计算服务（Nova），它表示云平台的工作负载的核心。如果有些云服务的工作中不包括计算，那么它们充其量只代表静态存储，但是所有动态活动都会涉及一些计算元素。OpenStack Compute 指的是一个特定的项目，该项目又称 Nova，OpenStack 的其他组件依托 Nova，与 Nova 协同工作，组成整个 OpenStack 云平台。从这里开始将正式进入云计算的世界。Nova，即计算服务，是 OpenStack 计算的弹性控制器。Nova 可以说是整个云平台最重要的组件，其功能包括运行虚拟机实例、管理网络，以及通过用户和项目来控制对云的访问。OpenStack 最基础的开源项目称为 Nova，它提供的软件可以控制基础即服务（IaaS）云计算平台，和 Amazon EC2、Rackspace 云服务器有一定程度相似。OpenStack Compute 没有包含任何的虚拟化软件，相反它定义和运行在主机操作系统上的虚拟化机制交互的驱动程序，并通过基于 Web 的程序应用接口(API)来提供功能的使用。

Nova-api 对外统一提供标准化接口，计算资源（Computer），存储资源（Volume）、网络资源（Network）等子模块通过相应的 API 接口服务对外提供服务。

API 接口操作数据库实现资源数据模型的维护。

通过消息中间件，通知相应的守护进程，如 nova-compute 实现服务接口等，API 与守护进程共享数据库，但守护进程侧重维护状态信息、网络资源状态等。

守护进程之间不能直接调用，需要通过 API 调用，如 nova-compute 为虚拟机分配网络，需要调用 network-api，而不能直接调用 nova-network，这样易于解耦合。

6.3.1 Nova 目录结构

Nova 目录结构如下：

Bin：Nova 服务执行文件。

Contrib：第三方贡献包。
Doc：技术文档。
Etc：Nova 配置文件样例。
Nova：Nova 代码目录。
Plugins：Nova 插件。
Smoketests：冒烟测试代码。
Tools：工具。
Babel.cfg：Flask-Babel 配置。
CONTRIBUTING.rst：贡献指南。
HACKING.rst：Hack 指南。
LICENSE：Apache2 LICENSE。
MANIFEST.in：打包规则。
OpenStack-common.conf：OSLO 配置。
Pylintrc：Pylint 代码分析配置。
README.rst：Nova 简介。
Run_tests.sh：测试案例。
Setup.cfg：setup.py 配置。
Setup.py：Nova 安装脚本。
Tox.ini：Python 的标准化测试。

6.3.2 安装和配置控制结点的计算服务

1. 在控制结点安装 nova 计算服务

在配置 OpenStack 网络（Neutron）服务时，用户必须创建一个数据库、服务凭证和 API 端点。

创建 nova 相关数据库，数据库名与密码一致（密码也可自行设置，牢记即可）。Nova 数据库创建命令如图 6-42~图 6-54 所示。

```
[root@controller ~]# mysql -u root -p123456
MariaDB [(none)]> CREATE DATABASE nova_api;
```

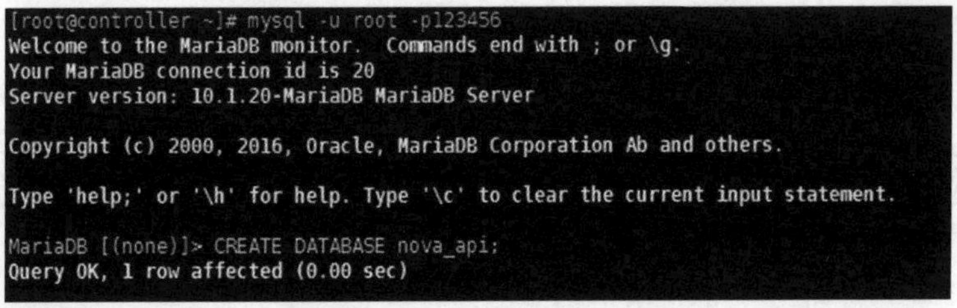

图 6-42 创建 nova 数据库 1

```
MariaDB [(none)]> CREATE DATABASE nova;
```

图 6-43 创建 nova 数据库 2

MariaDB [(none)]> CREATE DATABASE nova_cell0;

```
MariaDB [(none)]> CREATE DATABASE nova_cell0;
Query OK, 1 row affected (0.00 sec)
```

图 6-44　创建 nova 数据库 3

MariaDB [(none)]> CREATE DATABASE placement;

```
MariaDB [(none)]> CREATE DATABASE placement;
Query OK, 1 row affected (0.00 sec)
```

图 6-45　创建 nova 数据库 4

MariaDB [(none)]> GRANT ALL PRIVILEGES ON nova_api.* TO 'nova'@'localhost' IDENTIFIED BY 'nova';

```
MariaDB [(none)]> GRANT ALL PRIVILEGES ON nova_api.* TO 'nova'@'localhost' IDENTIFIED BY 'nova';
Query OK, 0 rows affected (0.00 sec)
```

图 6-46　创建 nova 数据库 5

MariaDB [(none)]> GRANT ALL PRIVILEGES ON nova_api.* TO 'nova'@'%' IDENTIFIED BY 'nova';

```
MariaDB [(none)]> GRANT ALL PRIVILEGES ON nova_api.* TO 'nova'@'%' IDENTIFIED BY 'nova';
Query OK, 0 rows affected (0.00 sec)
```

图 6-47　创建 nova 数据库 6

MariaDB [(none)]> GRANT ALL PRIVILEGES ON nova.* TO 'nova'@'localhost' IDENTIFIED BY 'nova';

```
MariaDB [(none)]> GRANT ALL PRIVILEGES ON nova.* TO 'nova'@'localhost' IDENTIFIED BY 'nova';
Query OK, 0 rows affected (0.00 sec)
```

图 6-48　创建 nova 数据库 7

MariaDB [(none)]> GRANT ALL PRIVILEGES ON nova.* TO 'nova'@'%' IDENTIFIED BY 'nova';

```
MariaDB [(none)]> GRANT ALL PRIVILEGES ON nova.* TO 'nova'@'%' IDENTIFIED BY 'nova';
Query OK, 0 rows affected (0.00 sec)
```

图 6-49　创建 nova 数据库 8

MariaDB [(none)]> GRANT ALL PRIVILEGES ON nova_cell0.* TO 'nova'@'localhost' IDENTIFIED BY 'nova';

```
MariaDB [(none)]> GRANT ALL PRIVILEGES ON nova_cell0.* TO 'nova'@'localhost' IDENTIFIED BY 'nova';
Query OK, 0 rows affected (0.00 sec)
```

图 6-50　创建 nova 数据库 9

单元 6　安装 OpenStack 服务

```
MariaDB [(none)]> GRANT ALL PRIVILEGES ON nova_cell0.* TO 'nova'@'%' IDENTIFIED BY 'nova';
```

```
MariaDB [(none)]> GRANT ALL PRIVILEGES ON nova_cell0.* TO 'nova'@'%' IDENTIFIED BY 'nova';
Query OK, 0 rows affected (0.00 sec)
```

图 6-51　创建 nova 数据库 10

```
MariaDB [(none)]> GRANT ALL PRIVILEGES ON placement.* TO 'placement'@'localhost' IDENTIFIED BY 'placement';
```

```
MariaDB [(none)]> GRANT ALL PRIVILEGES ON placement.* TO 'placement'@'localhost' IDENTIFIED BY 'placement';
Query OK, 0 rows affected (0.01 sec)
```

图 6-52　创建 nova 数据库 11

```
MariaDB [(none)]> GRANT ALL PRIVILEGES ON placement.* TO 'placement'@'%' IDENTIFIED BY 'placement';
```

```
MariaDB [(none)]> GRANT ALL PRIVILEGES ON placement.* TO 'placement'@'%' IDENTIFIED BY 'placement';
Query OK, 0 rows affected (0.00 sec)
```

图 6-53　创建 nova 数据库 12

```
MariaDB [(none)]> flush privileges;
```

```
MariaDB [(none)]> flush privileges;
Query OK, 0 rows affected (0.00 sec)
```

图 6-54　创建 nova 数据库 13

```
MariaDB [(none)]> exit
```

2. 在 Keystone 上面注册 nova 服务

创建服务证书

(1) 在 Keystone 上创建 nova 用户，如图 6-55 和图 6-56 所示。

```
# source keystone-admin-pass.sh
# openstack user create --domain default --password=nova nova
```

```
[root@controller ~]# source keystone-admin-pass.sh
[root@controller ~]# openstack user create --domain default --password=nova nova
+---------------------+----------------------------------+
| Field               | Value                            |
+---------------------+----------------------------------+
| domain_id           | default                          |
| enabled             | True                             |
| id                  | cf1ec6b675124537aebfce3cf98de251 |
| name                | nova                             |
| options             | {}                               |
| password_expires_at | None                             |
+---------------------+----------------------------------+
```

图 6-55　创建 nova 用户 1

```
# openstack user list
```

```
[root@controller ~]# openstack user list
+----------------------------------+---------+
| ID                               | Name    |
+----------------------------------+---------+
| 69a60187bc8449318ec3f72b919409bb | glance  |
| 8913182d285542eba893c857240bef4c | admin   |
| cf1ec6b675124537aebfce3cf98de251 | nova    |
| f18d2ad10e1b45a9986c8af3b0fbfc29 | myuser  |
+----------------------------------+---------+
```

图 6-56　创建 nova 用户 2

（2）在 Keystone 上将 nova 用户配置为 admin 角色并添加进 service 项目。

`# openstack role add --project service --user nova admin`

（3）创建 nova 计算服务的实体，如图 6-57 和图 6-58 所示。

`# openstack service create --name nova --description "OpenStack Compute" compute`

```
[root@controller ~]# openstack service create --name nova --description "OpenStack Compute" compute
+-------------+----------------------------------+
| Field       | Value                            |
+-------------+----------------------------------+
| description | OpenStack Compute                |
| enabled     | True                             |
| id          | 3925f1e709c74fbfa3dbdf4195826021 |
| name        | nova                             |
| type        | compute                          |
+-------------+----------------------------------+
```

图 6-57　创建 nova 计算服务实体 1

`# openstack service list`

```
[root@controller ~]# openstack service list
+----------------------------------+----------+----------+
| ID                               | Name     | Type     |
+----------------------------------+----------+----------+
| 3925f1e709c74fbfa3dbdf4195826021 | nova     | compute  |
| 7486a3b7749b41be96eb67e191be4baa | glance   | image    |
| c507b197de044ee7802e16077992561e | keystone | identity |
+----------------------------------+----------+----------+
```

图 6-58　创建 nova 计算服务实体 2

（4）创建计算服务的 API 端点（Endpoint），如图 6-59~图 6-62 所示。

`# openstack endpoint create --region RegionOne compute public http://controller:8774/v2.1`

```
[root@controller ~]# openstack endpoint create --region RegionOne compute public http://controller:8774/v2.1
openstack endpoint list+------------+----------------------------------+
| Field        | Value                            |
+--------------+----------------------------------+
| enabled      | True                             |
| id           | 9af286370a49459fa4841bc655feddle |
| interface    | public                           |
| region       | RegionOne                        |
| region_id    | RegionOne                        |
| service_id   | 3925f1e709c74fbfa3dbdf4195826021 |
| service_name | nova                             |
| service_type | compute                          |
| url          | http://controller:8774/v2.1      |
+--------------+----------------------------------+
```

图 6-59　创建计算服务 API 端点 1

openstack endpoint create --region RegionOne compute internal http://controller:8774/v2.1

图 6-60　创建计算服务 API 端点 2

openstack endpoint create --region RegionOne compute admin http://controller:8774/v2.1

图 6-61　创建计算服务 API 端点 3

openstack endpoint list

图 6-62　创建计算服务 API 端点 4

(5) 创建并注册 placement 项目的服务证书，如图 6-63~图 6-69 所示。

openstack user create --domain default --password=placement placement

```
[root@controller ~]# openstack user create --domain default --password=placement placement
+---------------------+----------------------------------+
| Field               | Value                            |
+---------------------+----------------------------------+
| domain_id           | default                          |
| enabled             | True                             |
| id                  | a9e8745a61ba498e8f406e8070427bf0 |
| name                | placement                        |
| options             | {}                               |
| password_expires_at | None                             |
+---------------------+----------------------------------+
```

图 6-63　创建并注册服务证书 1

```
# openstack role add --project service --user placement admin
```

```
[root@controller ~]# openstack role add --project service --user placement admin
[root@controller ~]#
```

图 6-64　创建并注册服务证书 2

```
# openstack service create --name placement --description "Placement API" placement
```

```
[root@controller ~]# openstack service create --name placement --description "Placement API" placement
+-------------+----------------------------------+
| Field       | Value                            |
+-------------+----------------------------------+
| description | Placement API                    |
| enabled     | True                             |
| id          | c7814063837b4bd7969ba3cd58fedb32 |
| name        | placement                        |
| type        | placement                        |
+-------------+----------------------------------+
```

图 6-65　创建并注册服务证书 3

创建 placement 项目的 endpoint（API 端口）。

```
# openstack endpoint create --region RegionOne placement public http://controller:8778
```

```
[root@controller ~]# openstack endpoint create --region RegionOne placement public http://controller:8778
+--------------+----------------------------------+
| Field        | Value                            |
+--------------+----------------------------------+
| enabled      | True                             |
| id           | ab4cc832f00c492a990fc09d0f0a6ec9 |
| interface    | public                           |
| region       | RegionOne                        |
| region_id    | RegionOne                        |
| service_id   | c7814063837b4bd7969ba3cd58fedb32 |
| service_name | placement                        |
| service_type | placement                        |
| url          | http://controller:8778           |
+--------------+----------------------------------+
```

图 6-66　创建并注册服务证书 4

```
# openstack endpoint create --region RegionOne placement internal http://controller:8778
```

单元 6 安装 OpenStack 服务

```
[root@controller ~]# openstack endpoint create --region RegionOne placement internal http://controller:8778
+--------------+----------------------------------+
| Field        | Value                            |
+--------------+----------------------------------+
| enabled      | True                             |
| id           | dbd166d8a192411780bf5d7909a76d57 |
| interface    | internal                         |
| region       | RegionOne                        |
| region_id    | RegionOne                        |
| service_id   | c7814063837b4bd7969ba3cd58fedb32 |
| service_name | placement                        |
| service_type | placement                        |
| url          | http://controller:8778           |
+--------------+----------------------------------+
```

图 6-67　创建并注册服务证书 5

openstack endpoint create --region RegionOne placement admin http://controller:8778

```
[root@controller ~]# openstack endpoint create --region RegionOne placement admin http://controller:8778
+--------------+----------------------------------+
| Field        | Value                            |
+--------------+----------------------------------+
| enabled      | True                             |
| id           | 548be49381d641e1a428cdbe09e48fa9 |
| interface    | admin                            |
| region       | RegionOne                        |
| region_id    | RegionOne                        |
| service_id   | c7814063837b4bd7969ba3cd58fedb32 |
| service_name | placement                        |
| service_type | placement                        |
| url          | http://controller:8778           |
+--------------+----------------------------------+
```

图 6-68　创建并注册服务证书 6

openstack endpoint list

```
[root@controller ~]# openstack endpoint list
+----------------------------------+-----------+--------------+--------------+---------+-----------+-----------------------------+
| ID                               | Region    | Service Name | Service Type | Enabled | Interface | URL                         |
+----------------------------------+-----------+--------------+--------------+---------+-----------+-----------------------------+
| 01ad3f54b25d4900b300248cb6a3515b | RegionOne | nova         | compute      | True    | admin     | http://controller:8774/v2.1 |
| 14167467f45947b8b796bec0c76637fa | RegionOne | nova         | compute      | True    | internal  | http://controller:8774/v2.1 |
| 1ade85b2725f49a3bb955df5b9f9b072 | RegionOne | keystone     | identity     | True    | public    | http://controller:5000/v3/  |
| 548be49381d641e1a428cdbe09e48fa9 | RegionOne | placement    | placement    | True    | admin     | http://controller:8778      |
| 7068d3761cdb4923a1f4bd93f44fdb29 | RegionOne | keystone     | identity     | True    | admin     | http://controller:5000/v3/  |
| 9af286370a49459fa4841bc655feddle | RegionOne | nova         | compute      | True    | public    | http://controller:8774/v2.1 |
| ab4cc832f00c492a990fc09d0f0a6ec9 | RegionOne | placement    | placement    | True    | public    | http://controller:8778      |
| c61260271feb40d3bd2250e5bfaa66ee | RegionOne | keystone     | identity     | True    | internal  | http://controller:5000/v3/  |
| cb1937f9b53647b286328afde2ddb0f4 | RegionOne | glance       | image        | True    | admin     | http://10.0.0.9:9292        |
| cb942271f26e42a1b83ea3ce42e0645b | RegionOne | glance       | image        | True    | public    | http://10.0.0.9:9292        |
| dbd166d8a192411780bf5d7909a76d57 | RegionOne | placement    | placement    | True    | internal  | http://controller:8778      |
| e4f806539567468ab77bce37ab7d4aa8 | RegionOne | glance       | image        | True    | internal  | http://10.0.0.9:9292        |
+----------------------------------+-----------+--------------+--------------+---------+-----------+-----------------------------+
```

图 6-69　创建并注册服务证书 7

3. 在控制结点安装 nova 相关服务

（1）安装 nova 相关软件包，如图 6-70 所示。

yum install openstack-nova-api openstack-nova-conductor \
  openstack-nova-console openstack-nova-novncproxy \
  openstack-nova-scheduler openstack-nova-placement-api -y

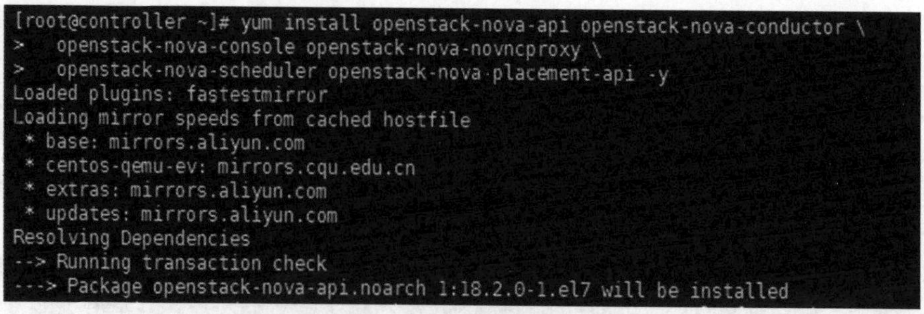

```
[root@controller ~]# yum install openstack-nova-api openstack-nova-conductor \
> openstack-nova-console openstack-nova-novncproxy \
> openstack-nova-scheduler openstack-nova-placement-api -y
Loaded plugins: fastestmirror
Loading mirror speeds from cached hostfile
 * base: mirrors.aliyun.com
 * centos-qemu-ev: mirrors.cqu.edu.cn
 * extras: mirrors.aliyun.com
 * updates: mirrors.aliyun.com
Resolving Dependencies
--> Running transaction check
---> Package openstack-nova-api.noarch 1:18.2.0-1.el7 will be installed
```

图 6-70　安装 nova 相关软件包

(2)快速修改 nova 配置，配置计算服务 Nova 使用 Keystone 进行认证，如图 6-71～图 6-74 所示。

```
# openstack-config --set  /etc/nova/nova.conf DEFAULT enabled_apis osapi_compute,metadata
# openstack-config --set  /etc/nova/nova.conf DEFAULT my_ip 10.0.0.9
# openstack-config --set  /etc/nova/nova.conf DEFAULT use_neutron true
# openstack-config --set  /etc/nova/nova.conf DEFAULT firewall_driver nova.virt.firewall.NoopFirewallDriver
# openstack-config --set  /etc/nova/nova.conf DEFAULT transport_url rabbit://openstack:openstack@controller
# openstack-config --set  /etc/nova/nova.conf api_database connection mysql+pymysql://nova:nova@controller/nova_api
# openstack-config --set  /etc/nova/nova.conf database connection mysql+pymysql://nova:nova@controller/nova
# openstack-config --set  /etc/nova/nova.conf placement_database connection mysql+pymysql://placement:placement@controller/placement
# openstack-config --set  /etc/nova/nova.conf api auth_strategy keystone
# openstack-config --set  /etc/nova/nova.conf keystone_authtoken auth_url http://controller:5000/v3
# openstack-config --set  /etc/nova/nova.conf keystone_authtoken memcached_servers controller:11211
# openstack-config --set  /etc/nova/nova.conf keystone_authtoken auth_type password
# openstack-config --set  /etc/nova/nova.conf keystone_authtoken project_domain_name default
# openstack-config --set  /etc/nova/nova.conf keystone_authtoken user_domain_name default
# openstack-config --set  /etc/nova/nova.conf keystone_authtoken project_name service
```

openstack-config --set /etc/nova/nova.conf keystone_authtoken username nova
openstack-config --set /etc/nova/nova.conf keystone_authtoken password nova
openstack-config --set /etc/nova/nova.conf vnc enabled true
openstack-config --set /etc/nova/nova.conf vnc server_listen '$my_ip'
openstack-config --set /etc/nova/nova.conf vnc server_proxyclient_address '$my_ip'
openstack-config --set /etc/nova/nova.conf glance api_servers http://controller:9292
openstack-config --set /etc/nova/nova.conf oslo_concurrency lock_path /var/lib/nova/tmp
openstack-config --set /etc/nova/nova.conf placement region_name RegionOne
openstack-config --set /etc/nova/nova.conf placement project_domain_name Default
openstack-config --set /etc/nova/nova.conf placement project_name service
openstack-config --set /etc/nova/nova.conf placement auth_type password
openstack-config --set /etc/nova/nova.conf placement user_domain_name Default
openstack-config --set /etc/nova/nova.conf placement auth_url http://controller:5000/v3
openstack-config --set /etc/nova/nova.conf placement username placement
openstack-config --set /etc/nova/nova.conf placement password placement
openstack-config --set /etc/nova/nova.conf scheduler discover_hosts_in_cells_interval 300

图 6-71 配置计算服务 1

```
[root@controller ~]# openstack-config --set /etc/nova/nova.conf placement auth_type password
[root@controller ~]# openstack-config --set /etc/nova/nova.conf placement user_domain_name Default
[root@controller ~]# openstack-config --set /etc/nova/nova.conf placement auth_url http://controller:5000/v3
[root@controller ~]# openstack-config --set /etc/nova/nova.conf placement username placement
[root@controller ~]# openstack-config --set /etc/nova/nova.conf placement password placement
[root@controller ~]# openstack-config --set /etc/nova/nova.conf scheduler discover_hosts_in_cells_interval 300
```

图 6-72 配置计算服务 2

检查生效的 nova 设置：

[root@controller ~]# egrep -v "^#|^$" /etc/nova/nova.conf

```
[root@controller ~]# egrep -v "^#|^$" /etc/nova/nova.conf
[DEFAULT]
enabled_apis = osapi_compute,metadata
my_ip = 10.0.0.9
use_neutron = true
firewall_driver = nova.virt.firewall.NoopFirewallDriver
transport_url = rabbit://openstack:openstack@controller
[api]
auth_strategy = keystone
```

图 6-73 配置计算服务 3

```
[vnc]
enabled = true
server_listen = $my_ip
server_proxyclient_address = $my_ip
[workarounds]
[wsgi]
[xenserver]
[xvp]
[zvm]
```

图 6-74 配置计算服务 4

(3) 修改 nova 的虚拟主机配置文件，如图 6-75 所示。

vim /etc/httpd/conf.d/00-nova-placement-api.conf

增加如下内容：

```
<Directory /usr/bin>
   <IfVersion >=2.4>
      Require all granted
   </IfVersion>
   <IfVersion < 2.4>
      Order allow, deny
      Allow from all
   </IfVersion>
</Directory>
```

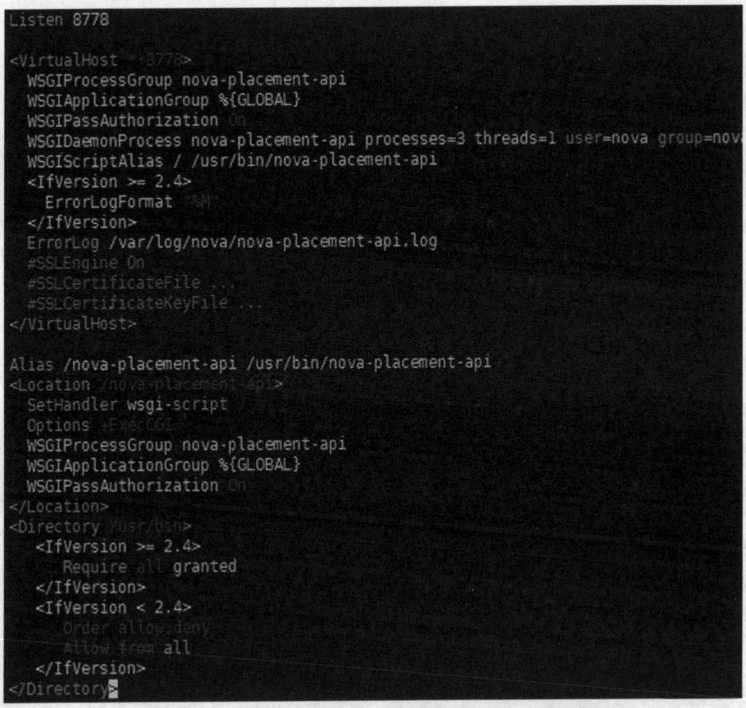

```
Listen 8778

<VirtualHost *:8778>
  WSGIProcessGroup nova-placement-api
  WSGIApplicationGroup %{GLOBAL}
  WSGIPassAuthorization On
  WSGIDaemonProcess nova-placement-api processes=3 threads=1 user=nova group=nov
  WSGIScriptAlias / /usr/bin/nova-placement-api
  <IfVersion >= 2.4>
    ErrorLogFormat "%M"
  </IfVersion>
  ErrorLog /var/log/nova/nova-placement-api.log
  #SSLEngine On
  #SSLCertificateFile ...
  #SSLCertificateKeyFile ...
</VirtualHost>

Alias /nova-placement-api /usr/bin/nova-placement-api
<Location /nova-placement-api>
  SetHandler wsgi-script
  Options +ExecCGI
  WSGIProcessGroup nova-placement-api
  WSGIApplicationGroup %{GLOBAL}
  WSGIPassAuthorization On
</Location>
<Directory /usr/bin>
   <IfVersion >= 2.4>
      Require all granted
   </IfVersion>
   <IfVersion < 2.4>
      Order allow,deny
      Allow from all
   </IfVersion>
</Directory>
```

图 6-75　修改 nova 虚拟主机配置文件

修改完后重启 httpd 服务，如图 6-76 所示。

```
# systemctl restart httpd
# systemctl status httpd
```

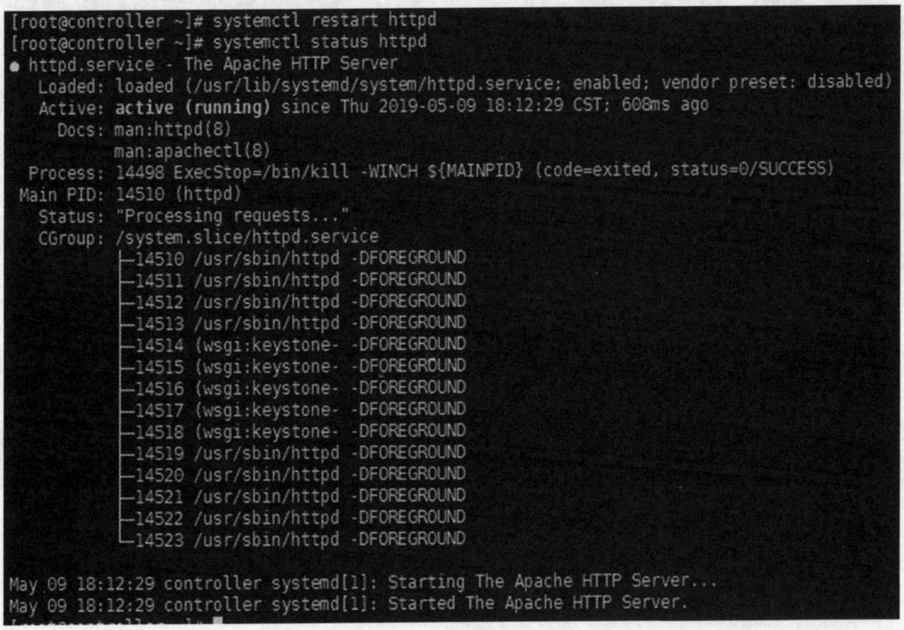

```
[root@controller ~]# systemctl restart httpd
[root@controller ~]# systemctl status httpd
● httpd.service - The Apache HTTP Server
   Loaded: loaded (/usr/lib/systemd/system/httpd.service; enabled; vendor preset: disabled)
   Active: active (running) since Thu 2019-05-09 18:12:29 CST; 608ms ago
     Docs: man:httpd(8)
           man:apachectl(8)
  Process: 14498 ExecStop=/bin/kill -WINCH ${MAINPID} (code=exited, status=0/SUCCESS)
 Main PID: 14510 (httpd)
   Status: "Processing requests..."
   CGroup: /system.slice/httpd.service
           ├─14510 /usr/sbin/httpd -DFOREGROUND
           ├─14511 /usr/sbin/httpd -DFOREGROUND
           ├─14512 /usr/sbin/httpd -DFOREGROUND
           ├─14513 /usr/sbin/httpd -DFOREGROUND
           ├─14514 (wsgi:keystone- -DFOREGROUND
           ├─14515 (wsgi:keystone- -DFOREGROUND
           ├─14516 (wsgi:keystone- -DFOREGROUND
           ├─14517 (wsgi:keystone- -DFOREGROUND
           ├─14518 (wsgi:keystone- -DFOREGROUND
           ├─14519 /usr/sbin/httpd -DFOREGROUND
           ├─14520 /usr/sbin/httpd -DFOREGROUND
           ├─14521 /usr/sbin/httpd -DFOREGROUND
           ├─14522 /usr/sbin/httpd -DFOREGROUND
           └─14523 /usr/sbin/httpd -DFOREGROUND

May 09 18:12:29 controller systemd[1]: Starting The Apache HTTP Server...
May 09 18:12:29 controller systemd[1]: Started The Apache HTTP Server.
```

图 6-76　重启 httpd 服务

4. 同步 nova 数据（注意同步顺序）

（1）初始化 nova-api 和 placement 数据库，如图 6-77~图 6-79 所示。

su -s /bin/sh -c "nova-manage api_db sync" nova

```
[root@controller ~]# su -s /bin/sh -c "nova-manage api_db sync" nova
[root@controller ~]#
```

图 6-77　同步 nova 数据 1

验证数据库：

mysql -h10.0.0.9 -unova -pnova -e "use nova_api;show tables;"

```
[root@controller ~]# mysql -h10.0.0.9 -unova -pnova -e "use nova_api;show tables;"
+-----------------------------+
| Tables_in_nova_api          |
+-----------------------------+
| aggregate_hosts             |
| aggregate_metadata          |
| aggregates                  |
| allocations                 |
| build_requests              |
| cell_mappings               |
| consumers                   |
| flavor_extra_specs          |
| flavor_projects             |
| flavors                     |
| host_mappings               |
| instance_group_member       |
| instance_group_policy       |
| instance_groups             |
| instance_mappings           |
| inventories                 |
| key_pairs                   |
| migrate_version             |
| placement_aggregates        |
| project_user_quotas         |
| projects                    |
| quota_classes               |
| quota_usages                |
| quotas                      |
| request_specs               |
| reservations                |
| resource_classes            |
| resource_provider_aggregates|
| resource_provider_traits    |
| resource_providers          |
| traits                      |
| users                       |
+-----------------------------+
```

图 6-78　同步 nova 数据 2

mysql -h10.0.0.9 -uplacement -pplacement -e "use placement;show tables;"

单元 6 安装 OpenStack 服务

```
[root@controller ~]# mysql -h10.0.0.9 -uplacement -pplacement -e "use placement;show tables;"
+-----------------------------+
| Tables_in_placement         |
+-----------------------------+
| aggregate_hosts             |
| aggregate_metadata          |
| aggregates                  |
| allocations                 |
| build_requests              |
| cell_mappings               |
| consumers                   |
| flavor_extra_specs          |
| flavor_projects             |
| flavors                     |
| host_mappings               |
| instance_group_member       |
| instance_group_policy       |
| instance_groups             |
| instance_mappings           |
| inventories                 |
| key_pairs                   |
| migrate_version             |
| placement_aggregates        |
| project_user_quotas         |
| projects                    |
| quota_classes               |
| quota_usages                |
| quotas                      |
| request_specs               |
| reservations                |
| resource_classes            |
| resource_provider_aggregates|
| resource_provider_traits    |
| resource_providers          |
| traits                      |
| users                       |
+-----------------------------+
```

图 6-79 同步 nova 数据 3

（2）初始化 nova_cell0 和 nova 数据库。

注册 cell0 数据库，如图 6-80 所示。

```
# su -s /bin/sh -c "nova-manage cell_v2 map_cell0" nova
```

```
[root@controller ~]# su -s /bin/sh -c "nova-manage cell_v2 map_cell0" nova
[root@controller ~]#
```

图 6-80 注册 cell0 数据库

创建 cell1 单元，如图 6-81 所示。

```
# su -s /bin/sh -c "nova-manage cell_v2 create_cell --name=cell1 --verbose" nova
```

```
[root@controller ~]# su -s /bin/sh -c "nova-manage cell_v2 create_cell --name=cell1 --verbose" nova
3b674c88-4036-4202-8572-edf9bfaa0065
[root@controller ~]#
```

图 6-81 创建 cell1 单元

初始化 nova 数据库，如图 6-82 所示。

```
# su -s /bin/sh -c "nova-manage db sync" nova
```

图 6-82 初始化 nova 数据库

可以看到上面有两个警告信息,不是很严重,后续版本会修复,再执行一次就不再报告。
验证数据库,如图 6-83 和图 6-84 所示。

```
# mysql -h10.0.0.9 -unova -pnova -e "use nova_cell0;show tables;"
```

图 6-83 验证数据库 1

```
# mysql -h10.0.0.9 -unova -pnova -e "use nova;show tables;"
```

图 6-84 验证数据库 2

(3)检查确认 cell0 和 cell1 注册是否成功,如图 6-85 所示。

```
# su -s /bin/sh -c "nova-manage cell_v2 list_cells" nova
```

单元 6 安装 OpenStack 服务

```
[root@controller ~]# su -s /bin/sh -c "nova-manage cell_v2 list_cells" nova
+-------+--------------------------------------+----------------------------------+--------------------------------------------+
| Name  | UUID                                 | Transport URL                    | Database Connection                        |
| Disabled |
+-------+--------------------------------------+----------------------------------+--------------------------------------------+
| cell0 | 00000000-0000-0000-0000-000000000000 | none:/                           | mysql+pymysql://nova:****@controller/nova_cel
l0 | False |
| cell1 | 3b674c88-4036-4202-8572-edf9bfaa0065 | rabbit://openstack:****@controller | mysql+pymysql://nova:****@controller/nova
| False |
+-------+--------------------------------------+----------------------------------+--------------------------------------------+
```

图 6-85　检查确认 cell0 和 cell1 注册是否成功

5. 启动 nova 服务

启动 nova 服务并设置为开机自启动，如图 6-86~图 6-93 所示。

```
# systemctl start openstack-nova-api.service openstack-nova-consoleauth.service \
    openstack-nova-scheduler.service openstack-nova-conductor.service \
    openstack-nova-novncproxy.service
```

```
[root@controller ~]# systemctl start openstack-nova-api.service openstack-nova-consoleauth.service \
>    openstack-nova-scheduler.service openstack-nova-conductor.service \
>    openstack-nova-novncproxy.service
[root@controller ~]#
```

图 6-86　启动 nova 服务 1

```
# systemctl status openstack-nova-api.service openstack-nova-consoleauth.service \
    openstack-nova-scheduler.service openstack-nova-conductor.service \
    openstack-nova-novncproxy.service
```

```
[root@controller ~]# systemctl status openstack-nova-api.service openstack-nova-consoleauth.service \
>    openstack-nova-scheduler.service openstack-nova-conductor.service \
>    openstack-nova-novncproxy.service
● openstack-nova-api.service - OpenStack Nova API Server
   Loaded: loaded (/usr/lib/systemd/system/openstack-nova-api.service; disabled; vendor preset: disabled)
   Active: active (running) since Thu 2019-05-09 19:01:22 CST; 24s ago
 Main PID: 16585 (nova-api)
   CGroup: /system.slice/openstack-nova-api.service
           ├─16585 /usr/bin/python2 /usr/bin/nova-api
           ├─16661 /usr/bin/python2 /usr/bin/nova-api
           ├─16662 /usr/bin/python2 /usr/bin/nova-api
           ├─16663 /usr/bin/python2 /usr/bin/nova-api
           ├─16664 /usr/bin/python2 /usr/bin/nova-api
           ├─16668 /usr/bin/python2 /usr/bin/nova-api
           ├─16669 /usr/bin/python2 /usr/bin/nova-api
           ├─16671 /usr/bin/python2 /usr/bin/nova-api
           └─16672 /usr/bin/python2 /usr/bin/nova-api

May 09 19:01:17 controller systemd[1]: Starting OpenStack Nova API Server...
May 09 19:01:22 controller systemd[1]: Started OpenStack Nova API Server.
```

图 6-87　启动 nova 服务 2

```
● openstack-nova-consoleauth.service - OpenStack Nova VNC console auth Server
   Loaded: loaded (/usr/lib/systemd/system/openstack-nova-consoleauth.service; disabled; vendor preset: disabled)
   Active: active (running) since Thu 2019-05-09 19:01:21 CST; 25s ago
 Main PID: 16586 (nova-consoleaut)
   CGroup: /system.slice/openstack-nova-consoleauth.service
           └─16586 /usr/bin/python2 /usr/bin/nova-consoleauth

May 09 19:01:17 controller systemd[1]: Starting OpenStack Nova VNC console auth Server...
May 09 19:01:21 controller systemd[1]: Started OpenStack Nova VNC console auth Server.
```

图 6-88 启动 nova 服务 3

```
● openstack-nova-scheduler.service - OpenStack Nova Scheduler Server
   Loaded: loaded (/usr/lib/systemd/system/openstack-nova-scheduler.service; disabled; vendor preset: disabled)
   Active: active (running) since Thu 2019-05-09 19:01:22 CST; 25s ago
 Main PID: 16587 (nova-scheduler)
   CGroup: /system.slice/openstack-nova-scheduler.service
           ├─16587 /usr/bin/python2 /usr/bin/nova-scheduler
           ├─16642 /usr/bin/python2 /usr/bin/nova-scheduler
           ├─16643 /usr/bin/python2 /usr/bin/nova-scheduler
           ├─16644 /usr/bin/python2 /usr/bin/nova-scheduler
           └─16645 /usr/bin/python2 /usr/bin/nova-scheduler

May 09 19:01:17 controller systemd[1]: Starting OpenStack Nova Scheduler Server...
May 09 19:01:22 controller systemd[1]: Started OpenStack Nova Scheduler Server.
```

图 6-89 启动 nova 服务 4

```
● openstack-nova-conductor.service - OpenStack Nova Conductor Server
   Loaded: loaded (/usr/lib/systemd/system/openstack-nova-conductor.service; disabled; vendor preset: disabled)
   Active: active (running) since Thu 2019-05-09 19:01:22 CST; 25s ago
 Main PID: 16588 (nova-conductor)
   CGroup: /system.slice/openstack-nova-conductor.service
           ├─16588 /usr/bin/python2 /usr/bin/nova-conductor
           ├─16651 /usr/bin/python2 /usr/bin/nova-conductor
           ├─16652 /usr/bin/python2 /usr/bin/nova-conductor
           ├─16653 /usr/bin/python2 /usr/bin/nova-conductor
           └─16654 /usr/bin/python2 /usr/bin/nova-conductor

May 09 19:01:17 controller systemd[1]: Starting OpenStack Nova Conductor Server...
May 09 19:01:22 controller systemd[1]: Started OpenStack Nova Conductor Server.
```

图 6-90 启动 nova 服务 5

```
● openstack-nova-novncproxy.service - OpenStack Nova NoVNC Proxy Server
   Loaded: loaded (/usr/lib/systemd/system/openstack-nova-novncproxy.service; disabled; vendor preset: disabled)
   Active: active (running) since Thu 2019-05-09 19:01:17 CST; 29s ago
 Main PID: 16589 (nova-novncproxy)
   CGroup: /system.slice/openstack-nova-novncproxy.service
           └─16589 /usr/bin/python2 /usr/bin/nova-novncproxy --web /usr/share/novnc/

May 09 19:01:17 controller systemd[1]: Started OpenStack Nova NoVNC Proxy Server.
```

图 6-91 启动 nova 服务 6

```
# systemctl enable openstack-nova-api.service openstack-nova-consoleauth.service \
    openstack-nova-scheduler.service openstack-nova-conductor.service \
    openstack-nova-novncproxy.service
```

图 6-92 启动 nova 服务 7

```
# systemctl list-unit-files |grep openstack-nova* |grep enabled
```

图 6-93 启动 nova 服务 8

6. 使用 Linux KVM 安装计算服务

计算结点根据从控制结点接收的请求运行虚拟机，计算服务依靠虚拟化引擎运行虚拟机，OpenStack 可以使用多种虚拟化引擎，这里使用 Linux KVM。

（1）安装计算服务，如图 6-94 所示。

```
[root@compute ~]# yum install openstack-nova-compute python-openstackclient openstack-utils -y
```

图 6-94 安装计算服务

（2）快速修改配置文件（/etc/nova/nova.conf），设为 Nova 配置数据库的位置，并配置 Nova 使用 Keystone 进行认证，如图 6-95 所示。

```
#openstack-config --set  /etc/nova/nova.conf DEFAULT my_ip 10.0.0.8
#openstack-config --set  /etc/nova/nova.conf DEFAULT use_neutron True
#openstack-config --set  /etc/nova/nova.conf DEFAULT firewall_driver nova.virt.firewall.NoopFirewallDriver
#openstack-config --set  /etc/nova/nova.conf DEFAULT enabled_apis osapi_compute,metadata
#openstack-config --set  /etc/nova/nova.conf DEFAULT transport_url rabbit://openstack:openstack@controller
```

```
#openstack-config --set  /etc/nova/nova.conf api auth_strategy keystone
#openstack-config --set  /etc/nova/nova.conf keystone_authtoken auth_url http://controller:5000/v3
#openstack-config --set  /etc/nova/nova.conf keystone_authtoken memcached_servers controller:11211
#openstack-config --set  /etc/nova/nova.conf keystone_authtoken auth_type password
#openstack-config --set  /etc/nova/nova.conf keystone_authtoken project_domain_name default
#openstack-config --set  /etc/nova/nova.conf keystone_authtoken user_domain_name default
#openstack-config --set  /etc/nova/nova.conf keystone_authtoken project_name service
#openstack-config --set  /etc/nova/nova.conf keystone_authtoken username nova
#openstack-config --set  /etc/nova/nova.conf keystone_authtoken password nova
#openstack-config --set  /etc/nova/nova.conf vnc enabled True
#openstack-config --set  /etc/nova/nova.conf vnc server_listen 0.0.0.0
#openstack-config --set  /etc/nova/nova.conf vnc server_proxyclient_address '$my_ip'
#openstack-config --set  /etc/nova/nova.conf vnc novncproxy_base_url http://controller:6080/vnc_auto.html
#openstack-config --set  /etc/nova/nova.conf glance api_servers http://controller:9292
#openstack-config --set  /etc/nova/nova.conf oslo_concurrency lock_path /var/lib/nova/tmp
#openstack-config --set  /etc/nova/nova.conf placement region_name RegionOne
#openstack-config --set  /etc/nova/nova.conf placement project_domain_name Default
#openstack-config --set  /etc/nova/nova.conf placement project_name service
#openstack-config --set  /etc/nova/nova.conf placement auth_type password
#openstack-config --set  /etc/nova/nova.conf placement user_domain_name Default
```

#openstack-config --set　/etc/nova/nova.conf placement auth_url http://controller:5000/v3
　　#openstack-config --set　/etc/nova/nova.conf placement username placement
　　#openstack-config --set　/etc/nova/nova.conf placement password placement

```
[root@compute ~]# openstack-config --set /etc/nova/nova.conf placement project_domain_name Default
[root@compute ~]# openstack-config --set /etc/nova/nova.conf placement project_name service
[root@compute ~]# openstack-config --set /etc/nova/nova.conf placement auth_type password
[root@compute ~]# openstack-config --set /etc/nova/nova.conf placement user_domain_name Default
[root@compute ~]# openstack-config --set /etc/nova/nova.conf placement auth_url http://controller:5000/v3
[root@compute ~]# openstack-config --set /etc/nova/nova.conf placement username placement
[root@compute ~]# openstack-config --set /etc/nova/nova.conf placement password placement
```

图 6-95　修改配置文件

查看生效的配置：
egrep -v "^#|^$" /etc/nova/nova.conf
（3）配置虚拟机的硬件加速。
首先确定计算结点是否支持虚拟机的硬件加速。
egrep -c '(vmx|svm)' /proc/cpuinfo
如果返回为 0，表示计算结点不支持硬件加速，需要配置 libvirt 使用 QEMU 方式管理虚拟机，使用以下命令，如图 6-96 所示。
openstack-config --set　/etc/nova/nova.conf libvirt virt_type　qemu
egrep -v "^#|^$" /etc/nova/nova.conf|grep 'virt_type'

```
[root@compute ~]# openstack-config --set /etc/nova/nova.conf libvirt virt_type qemu
[root@compute ~]# egrep -v "^#|^$" /etc/nova/nova.conf|grep 'virt_type'
virt_type = qemu
[root@compute ~]#
```

图 6-96　配置虚拟机

如果返回为其他值，表示计算结点支持硬件加速且不需要额外的配置，使用以下命令：
openstack-config --set　/etc/nova/nova.conf libvirt virt_type　kvm
egrep -v "^#|^$" /etc/nova/nova.conf|grep 'virt_type'
（4）启动 nova 相关服务，并配置为开机自启动，如图 6-97~图 6-99 所示。
systemctl start libvirtd.service openstack-nova-compute.service
#systemctl status libvirtd.service openstack-nova-compute.service

```
[root@compute ~]# systemctl start libvirtd.service openstack-nova-compute.service
st-unit-files |grep openstack-nova-compute.service[root@compute ~]# systemctl status libvirtd.service openstack-nova-compute.servic
e
● libvirtd.service - Virtualization daemon
   Loaded: loaded (/usr/lib/systemd/system/libvirtd.service; enabled; vendor preset: enabled)
   Active: active (running) since Fri 2019-05-10 10:19:38 CST; 3s ago
     Docs: man:libvirtd(8)
           https://libvirt.org
 Main PID: 8043 (libvirtd)
   CGroup: /system.slice/libvirtd.service
           └─8043 /usr/sbin/libvirtd
```

图 6-97　启动 nova 相关服务 1

```
● openstack-nova-compute.service - OpenStack Nova Compute Server
   Loaded: loaded (/usr/lib/systemd/system/openstack-nova-compute.service; disabled; vendor preset: disabled)
   Active: active (running) since Fri 2019-05-10 10:19:41 CST; 17ms ago
 Main PID: 8060 (nova-compute)
   CGroup: /system.slice/openstack-nova-compute.service
           └─8060 /usr/bin/python2 /usr/bin/nova-compute

May 10 10:19:38 compute systemd[1]: Starting OpenStack Nova Compute Server...
May 10 10:19:41 compute systemd[1]: Started OpenStack Nova Compute Server.
Hint: Some lines were ellipsized, use -l to show in full.
```

图 6-98　启动 nova 相关服务 2

```
# systemctl enable libvirtd.service openstack-nova-compute.service
# systemctl list-unit-files |grep libvirtd.service
# systemctl list-unit-files |grep openstack-nova-compute.service
```

```
[root@compute ~]# systemctl list-unit-files |grep libvirtd.service
libvirtd.service                              enabled
[root@compute ~]# systemctl list-unit-files |grep openstack-nova-compute.service
openstack-nova-compute.service                enabled
[root@compute ~]#
```

图 6-99　启动 nova 相关服务 3

（5）将计算结点增加到 cell 数据库，以下命令在控制结点操作，如图 6-100 所示。

[root@controller ~]# source keystone-admin-pass.sh

```
[root@controller ~]# source keystone-admin-pass.sh
[root@controller ~]#
```

图 6-100　增加到 cell 数据库

检查确认数据库有新的计算结点，如图 6-101 所示。

[root@controller ~]# openstack compute service list --service nova-compute

```
[root@controller ~]# openstack compute service list --service nova-compute
+----+--------------+---------+------+---------+-------+----------------------------+
| ID | Binary       | Host    | Zone | Status  | State | Updated At                 |
+----+--------------+---------+------+---------+-------+----------------------------+
| 10 | nova-compute | compute | nova | enabled | up    | 2019-05-10T02:24:05.000000 |
+----+--------------+---------+------+---------+-------+----------------------------+
[root@controller ~]#
```

图 6-101　确认新计算结点

手动将新的计算结点添加到 OpenStack 集群，如图 6-102 所示。

[root@controller ~]# su -s /bin/sh -c "nova-manage cell_v2 discover_hosts --verbose" nova

```
[root@controller ~]# su -s /bin/sh -c "nova-manage cell_v2 discover_hosts --verbose" nova
Found 2 cell mappings.
Skipping cell0 since it does not contain hosts.
Getting computes from cell 'cell1': 3b674c88-4036-4202-8572-edf9bfaa0065
Found 0 unmapped computes in cell: 3b674c88-4036-4202-8572-edf9bfaa0065
[root@controller ~]#
```

图 6-102　手动添加新计算结点

6.3.3 验证控制结点与计算结点的计算服务

在控制结点进行以下验证：

(1) 应用管理员环境脚本，如图 6-103 所示。

`[root@controller ~]# source keystone-admin-pass.sh`

```
[root@controller ~]# source keystone-admin-pass.sh
[root@controller ~]#
```

图 6-103 应用管理员环境脚本

(2) 列表查看安装 nova 服务组件，验证是否成功注册并启动了每个进程，如图 6-104 所示。

`[root@controller ~]# openstack compute service list`

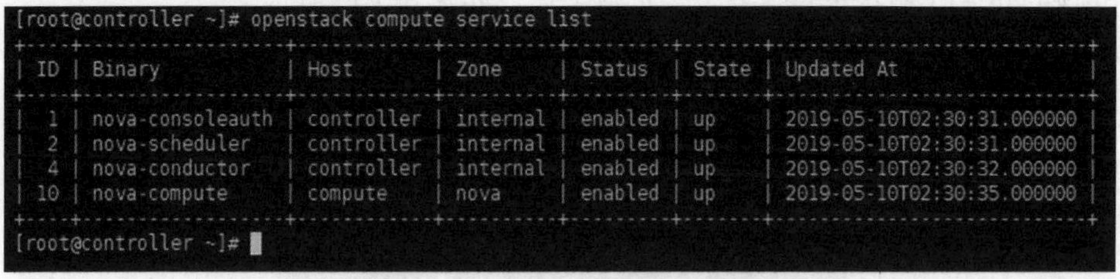

```
[root@controller ~]# openstack compute service list
+----+------------------+------------+----------+---------+-------+----------------------------+
| ID | Binary           | Host       | Zone     | Status  | State | Updated At                 |
+----+------------------+------------+----------+---------+-------+----------------------------+
|  1 | nova-consoleauth | controller | internal | enabled | up    | 2019-05-10T02:30:31.000000 |
|  2 | nova-scheduler   | controller | internal | enabled | up    | 2019-05-10T02:30:31.000000 |
|  4 | nova-conductor   | controller | internal | enabled | up    | 2019-05-10T02:30:32.000000 |
| 10 | nova-compute     | compute    | nova     | enabled | up    | 2019-05-10T02:30:35.000000 |
+----+------------------+------------+----------+---------+-------+----------------------------+
[root@controller ~]#
```

图 6-104 查看安装 nova 服务组件

(3) 在身份认证服务中列出 API 端点以验证其连接性，如图 6-105 所示。

`[root@controller ~]# openstack catalog list`

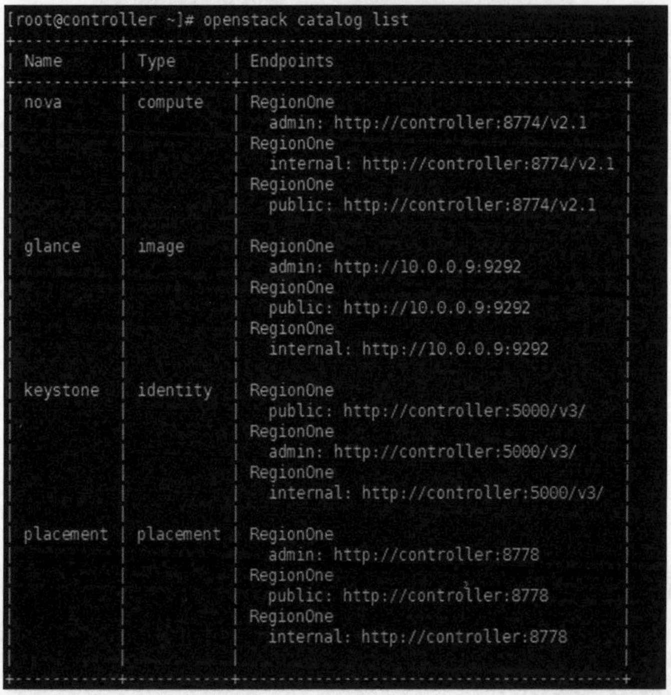

```
[root@controller ~]# openstack catalog list
+-----------+-----------+-------------------------------------------+
| Name      | Type      | Endpoints                                 |
+-----------+-----------+-------------------------------------------+
| nova      | compute   | RegionOne                                 |
|           |           |   admin: http://controller:8774/v2.1      |
|           |           | RegionOne                                 |
|           |           |   internal: http://controller:8774/v2.1   |
|           |           | RegionOne                                 |
|           |           |   public: http://controller:8774/v2.1     |
|           |           |                                           |
| glance    | image     | RegionOne                                 |
|           |           |   admin: http://10.0.0.9:9292             |
|           |           | RegionOne                                 |
|           |           |   public: http://10.0.0.9:9292            |
|           |           | RegionOne                                 |
|           |           |   internal: http://10.0.0.9:9292          |
|           |           |                                           |
| keystone  | identity  | RegionOne                                 |
|           |           |   public: http://controller:5000/v3/      |
|           |           | RegionOne                                 |
|           |           |   admin: http://controller:5000/v3/       |
|           |           | RegionOne                                 |
|           |           |   internal: http://controller:5000/v3/    |
|           |           |                                           |
| placement | placement | RegionOne                                 |
|           |           |   admin: http://controller:8778           |
|           |           | RegionOne                                 |
|           |           |   public: http://controller:8778          |
|           |           | RegionOne                                 |
|           |           |   internal: http://controller:8778        |
+-----------+-----------+-------------------------------------------+
```

图 6-105 列出 API 端点

(4) 在镜像服务中列出已有镜像以检查镜像服务的连接性, 如图 6-106 所示。

```
[root@controller ~]# openstack image list
```

图 6-106 列出已有镜像

(5) 检查 nova 各组件的状态。

检查 placement API 和 cell 的服务是否正常工作, 如图 6-107 所示。

```
[root@controller ~]# nova-status upgrade check
```

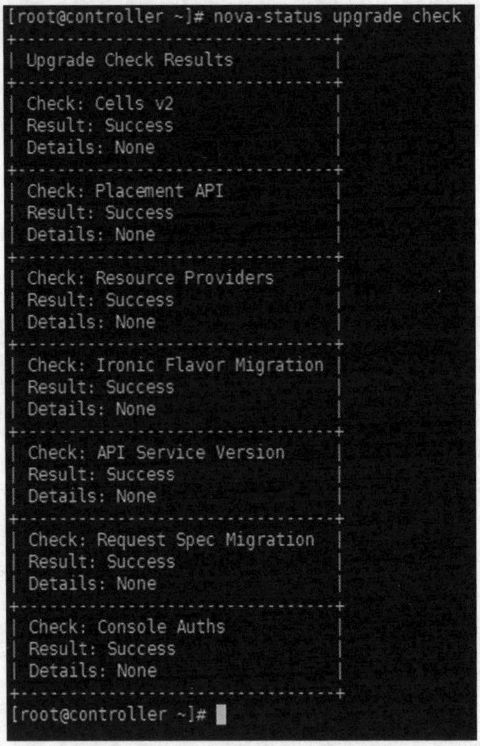

图 6-107 检查各组件状态

6.4 网络部署服务 Neutron 的安装配置

6.4.1 Neutron 目录结构

Neutron 目录结构如下：

Bin：Neutron 服务执行文件。
Contrib：第三方贡献包。
Doc：技术文档。
Etc：Neutron 配置文件样例。
Quantum：Neutron 代码目录。
Tools：工具。
Babel.cfg：Flask-Babel 配置。
HACKING.rst：Hack 指南。
LICENSE：Apache2 LICENSE。
MANIFEST.in：打包规则。
OpenStack-common.conf：OSLO 配置。
README：Neutron 简介。
Run_tests.py：测试案例。
Run_tests.sh：测试案例。
Setup.cfg：setup.py 配置。
Setup.py：Neutron 安装脚本。
TESTING：测试方法简介。
Tox.ini：Python 的标准化测试。

6.4.2 安装和配置控制结点网络服务

1. 创建 Neutron 数据库

在控制结点创建 Neutron 数据库，首先进入数据库执行 MySQL 命令后输入数据库密码进入数据库。

```
# mysql -p123456
```

创建 Neutron 数据库，并绑定权限设置密码（此次安装数据库密码设置为 neutron）

```
mysql> CREATE DATABASE neutron;
mysql> GRANT ALL PRIVILEGES ON neutron.* TO 'neutron'@'localhost' IDENTIFIED BY 'neutron';
mysql> GRANT ALL PRIVILEGES ON neutron.* TO 'neutron'@'%' IDENTIFIED BY 'neutron';
mysql> flush privileges;
mysql> exit
```

2. 在 Keystone 上创建 neutron 用户

在控制结点执行以下命令完成操作，如图 6-108 和图 6-109 所示。

```
[root@controller ~]# source keystone-admin-pass.sh
[root@controller ~]# openstack user create --domain default --password=neutron neutron
```

```
[root@controller ~]# source keystone-admin-pass.sh
[root@controller ~]# openstack user create --domain default --password=neutron neutron
+---------------------+----------------------------------+
| Field               | Value                            |
+---------------------+----------------------------------+
| domain_id           | default                          |
| enabled             | True                             |
| id                  | 1a8b73e9e9d94075b647529a6128b485 |
| name                | neutron                          |
| options             | {}                               |
| password_expires_at | None                             |
+---------------------+----------------------------------+
```

图 6-108　创建 neutron 用户 1

[root@controller ~]# openstack user list

```
[root@controller ~]# openstack user list
+----------------------------------+-----------+
| ID                               | Name      |
+----------------------------------+-----------+
| 1a8b73e9e9d94075b647529a6128b485 | neutron   |
| 69a60187bc8449318ec3f72b919409bb | glance    |
| 8913182d285542eba893c857240bef4c | admin     |
| a9e8745a61ba498e8f406e8070427bf0 | placement |
| cf1ec6b675124537aebfce3cf98de251 | nova      |
| f18d2ad10e1b45a9986c8af3b0fbfc29 | myuser    |
+----------------------------------+-----------+
```

图 6-109　创建 neutron 用户 2

3. 将 neutron 添加到 service 项目并授予 admin 角色（见图 6-110）

[root@controller ~]# openstack role add --project service --user neutron admin

```
[root@controller ~]# openstack role add --project service --user neutron admin
[root@controller ~]#
```

图 6-110　授予 admin 角色

4. 创建 neutron 服务实体

在控制结点执行以下命令完成安装软件，如图 6-111 和图 6-112 所示。

[root@controller ~]# openstack service create --name neutron --description "OpenStack Networking" network

```
[root@controller ~]# openstack service create --name neutron --description "OpenStack Networking" network
+-------------+----------------------------------+
| Field       | Value                            |
+-------------+----------------------------------+
| description | OpenStack Networking             |
| enabled     | True                             |
| id          | 97b7ed912ac5433dbaad47953fa1fac4 |
| name        | neutron                          |
| type        | network                          |
+-------------+----------------------------------+
```

图 6-111　创建 neutron 服务实体 1

[root@controller ~]# openstack service list

```
[root@controller ~]# openstack service list
+----------------------------------+-----------+-----------+
| ID                               | Name      | Type      |
+----------------------------------+-----------+-----------+
| 3925f1e709c74fbfa3dbdf4195826021 | nova      | compute   |
| 7486a3b7749b41be96eb67e191be4baa | glance    | image     |
| 97b7ed912ac5433dbaad47953fa1fac4 | neutron   | network   |
| c507b197de044ee7802e16077992561e | keystone  | identity  |
| c7814063837b4bd7969ba3cd58fedb32 | placement | placement |
+----------------------------------+-----------+-----------+
[root@controller ~]#
```

图 6-112　创建 neutron 服务实体 2

5. 创建 neutron 网络服务的 API 端点（Endpoint）

命令如图 6-113~图 6-116 所示。

[root@controller ~]# openstack endpoint create --region RegionOne network public http://controller:9696

```
[root@controller ~]# openstack endpoint create --region RegionOne network public http://controller:9696
+--------------+----------------------------------+
| Field        | Value                            |
+--------------+----------------------------------+
| enabled      | True                             |
| id           | d40c9f5fd2324c9983befad0a6e290cc |
| interface    | public                           |
| region       | RegionOne                        |
| region_id    | RegionOne                        |
| service_id   | 97b7ed912ac5433dbaad47953fa1fac4 |
| service_name | neutron                          |
| service_type | network                          |
| url          | http://controller:9696           |
+--------------+----------------------------------+
[root@controller ~]#
```

图 6-113　创建 neutron 网络服务 API 端点 1

[root@controller ~]# openstack endpoint create --region RegionOne network internal http://controller:9696

```
[root@controller ~]# openstack endpoint create --region RegionOne network internal http://controller:9696
+--------------+----------------------------------+
| Field        | Value                            |
+--------------+----------------------------------+
| enabled      | True                             |
| id           | d7ee8f021439403db2d0261169311e6f |
| interface    | internal                         |
| region       | RegionOne                        |
| region_id    | RegionOne                        |
| service_id   | 97b7ed912ac5433dbaad47953fa1fac4 |
| service_name | neutron                          |
| service_type | network                          |
| url          | http://controller:9696           |
+--------------+----------------------------------+
[root@controller ~]#
```

图 6-114　创建 neutron 网络服务 API 端点 2

[root@controller ~]# openstack endpoint create --region RegionOne network admin http://controller:9696

```
[root@controller ~]# openstack endpoint create --region RegionOne network admin http://controller:9696
+-------------+----------------------------------+
| Field       | Value                            |
+-------------+----------------------------------+
| enabled     | True                             |
| id          | a5e20a7bec2549bd9d6482851f17a57c |
| interface   | admin                            |
| region      | RegionOne                        |
| region_id   | RegionOne                        |
| service_id  | 97b7ed912ac5433dbaad47953fa1fac4 |
| service_name| neutron                          |
| service_type| network                          |
| url         | http://controller:9696           |
+-------------+----------------------------------+
[root@controller ~]#
```

图 6-115　创建 neutron 网络服务 API 端点 3

`[root@controller ~]# openstack endpoint list`

```
[root@controller ~]# openstack endpoint list
+----------------------------------+-----------+--------------+--------------+---------+-----------+-----------------------------+
| ID                               | Region    | Service Name | Service Type | Enabled | Interface | URL                         |
+----------------------------------+-----------+--------------+--------------+---------+-----------+-----------------------------+
| 01ad3f54b25d4900b300248cb6a3515b | RegionOne | nova         | compute      | True    | admin     | http://controller:8774/v2.1 |
| 14167467f45947b8b796bec0c76637fa | RegionOne | nova         | compute      | True    | internal  | http://controller:8774/v2.1 |
| 1ade85b2725f49a3bb955df5b9f9b072 | RegionOne | keystone     | identity     | True    | public    | http://controller:5000/v3/  |
| 548be49381d641e1a428cdbe09e48fa9 | RegionOne | placement    | placement    | True    | admin     | http://controller:8778      |
| 7068d3761cdb4923a1f4bd93f44fdb29 | RegionOne | keystone     | identity     | True    | admin     | http://controller:5000/v3/  |
| 9af286370a49459fa4841bc655fedd1e | RegionOne | nova         | compute      | True    | public    | http://controller:8774/v2.1 |
| a5e20a7bec2549bd9d6482851f17a57c | RegionOne | neutron      | network      | True    | admin     | http://controller:9696      |
| ab4cc832f00c492a990fc09d0f0a6ec9 | RegionOne | placement    | placement    | True    | public    | http://controller:8778      |
| c61260271feb40d3bd2250e5bfaa66ee | RegionOne | keystone     | identity     | True    | internal  | http://controller:5000/v3/  |
| cb1937f9b53647b286328afde2ddb0f4 | RegionOne | glance       | image        | True    | admin     | http://10.0.0.9:9292        |
| cb942271f26e42a1b83ea3ce42e0645b | RegionOne | glance       | image        | True    | public    | http://10.0.0.9:9292        |
| d40c9f5fd2324c9983befad0a6e290cc | RegionOne | neutron      | network      | True    | public    | http://controller:9696      |
| d7ee8f021439403db2d0261169311e6f | RegionOne | neutron      | network      | True    | internal  | http://controller:9696      |
| dbd166d8a192411780bf5d7909a76d57 | RegionOne | placement    | placement    | True    | internal  | http://controller:8778      |
| e4f806539567468ab77bce37ab7d4aa8 | RegionOne | glance       | image        | True    | internal  | http://10.0.0.9:9292        |
+----------------------------------+-----------+--------------+--------------+---------+-----------+-----------------------------+
[root@controller ~]#
```

图 6-116　创建 neutron 网络服务 API 端点 4

6. 在控制结点安装 neutron 网络组件

（1）安装 neutron 软件包，如图 6-117 所示。

`[root@controller ~]# yum install openstack-neutron openstack-neutron-ml2 openstack-neutron-linuxbridge ebtables -y`

```
[root@controller ~]# yum install openstack-neutron openstack-neutron-ml2 openstack-neutron-linuxbridge ebtables -y
Loaded plugins: fastestmirror
Loading mirror speeds from cached hostfile
 * base: mirrors.aliyun.com
 * centos-qemu-ev: mirrors.cqu.edu.cn
 * extras: mirrors.aliyun.com
 * updates: mirrors.aliyun.com
base                                                           | 3.6 kB  00:00:00
centos-ceph-luminous                                           | 2.9 kB  00:00:00
centos-openstack-rocky                                         | 2.9 kB  00:00:00
centos-qemu-ev                                                 | 2.9 kB  00:00:00
epel                                                           | 4.7 kB  00:00:00
extras                                                         | 3.4 kB  00:00:00
updates                                                        | 3.4 kB  00:00:00
(1/2): epel/x86_64/updateinfo                                  | 995 kB  00:00:00
(2/2): epel/x86_64/primary_db                                  | 6.7 MB  00:00:00
```

图 6-117　安装 neutron 软件包

单元 6　安装 OpenStack 服务

（2）快速配置 /etc/neutron/neutron.conf。

```
# openstack-config --set /etc/neutron/neutron.conf database connection mysql+pymysql://neutron:neutron@controller/neutron
# openstack-config --set /etc/neutron/neutron.conf DEFAULT core_plugin ml2
# openstack-config --set /etc/neutron/neutron.conf DEFAULT service_plugins
# openstack-config --set /etc/neutron/neutron.conf DEFAULT transport_url rabbit://openstack:openstack@controller
# openstack-config --set /etc/neutron/neutron.conf DEFAULT auth_strategy keystone
# openstack-config --set /etc/neutron/neutron.conf keystone_authtoken www_authenticate_uri http://controller:5000
# openstack-config --set /etc/neutron/neutron.conf keystone_authtoken auth_url http://controller:5000
# openstack-config --set /etc/neutron/neutron.conf keystone_authtoken memcached_servers controller:11211
# openstack-config --set /etc/neutron/neutron.conf keystone_authtoken auth_type password
# openstack-config --set /etc/neutron/neutron.conf keystone_authtoken project_domain_name default
# openstack-config --set /etc/neutron/neutron.conf keystone_authtoken user_domain_name default
# openstack-config --set /etc/neutron/neutron.conf keystone_authtoken project_name service
# openstack-config --set /etc/neutron/neutron.conf keystone_authtoken username neutron
# openstack-config --set /etc/neutron/neutron.conf keystone_authtoken password neutron
# openstack-config --set /etc/neutron/neutron.conf DEFAULT notify_nova_on_port_status_changes True
# openstack-config --set /etc/neutron/neutron.conf DEFAULT notify_nova_on_port_data_changes True
# openstack-config --set /etc/neutron/neutron.conf nova auth_url http://controller:5000
# openstack-config --set /etc/neutron/neutron.conf nova auth_type password
# openstack-config --set /etc/neutron/neutron.conf nova project_
```

domain_name default
　　# openstack-config --set /etc/neutron/neutron.conf nova user_domain_name default
　　# openstack-config --set /etc/neutron/neutron.conf nova region_name RegionOne
　　# openstack-config --set /etc/neutron/neutron.conf nova project_name service
　　# openstack-config --set /etc/neutron/neutron.conf nova username nova
　　# openstack-config --set /etc/neutron/neutron.conf nova password nova
　　# openstack-config --set /etc/neutron/neutron.conf oslo_concurrency lock_path /var/lib/neutron/tmp

查看生效的配置，如图 6-118 所示。

[root@controller ~]# egrep -v '(^$|^#)' /etc/neutron/neutron.conf

```
[root@controller ~]# egrep -v '(^$|^#)' /etc/neutron/neutron.conf
[DEFAULT]
core_plugin = ml2
service_plugins =
transport_url = rabbit://openstack:openstack@controller
auth_strategy = keystone
notify_nova_on_port_status_changes = True
notify_nova_on_port_data_changes = True
[agent]
[cors]
[database]
connection = mysql+pymysql://neutron:neutron@controller/neutron
[keystone_authtoken]
www_authenticate_uri = http://controller:5000
auth_url = http://controller:5000
memcached_servers = controller:11211
auth_type = password
project_domain_name = default
user_domain_name = default
project_name = service
username = neutron
password = neutron
[matchmaker_redis]
[nova]
auth_url = http://controller:5000
auth_type = password
project_domain_name = default
user_domain_name = default
region_name = RegionOne
project_name = service
username = nova
password = nova
[oslo_concurrency]
lock_path = /var/lib/neutron/tmp
[oslo_messaging_amqp]
[oslo_messaging_kafka]
[oslo_messaging_notifications]
[oslo_messaging_rabbit]
[oslo_messaging_zmq]
[oslo_middleware]
[oslo_policy]
[quotas]
[ssl]
[root@controller ~]#
```

图 6-118　查看生效配置 1

(3) 快速配置 /etc/neutron/plugins/ml2/ml2_conf.ini。

```
# openstack-config --set /etc/neutron/plugins/ml2/ml2_conf.ini ml2 type_drivers flat,vlan
# openstack-config --set /etc/neutron/plugins/ml2/ml2_conf.ini ml2 tenant_network_types
# openstack-config --set /etc/neutron/plugins/ml2/ml2_conf.ini ml2 mechanism_drivers linuxbridge
# openstack-config --set /etc/neutron/plugins/ml2/ml2_conf.ini ml2 extension_drivers port_security
# openstack-config --set /etc/neutron/plugins/ml2/ml2_conf.ini ml2_type_flat flat_networks provider
# openstack-config --set /etc/neutron/plugins/ml2/ml2_conf.ini securitygroup enable_ipset True
```

查看生效的配置，如图 6-119 所示。

```
[root@controller ~]# egrep -v '(^$|^#)' /etc/neutron/plugins/ml2/ml2_conf.ini
```

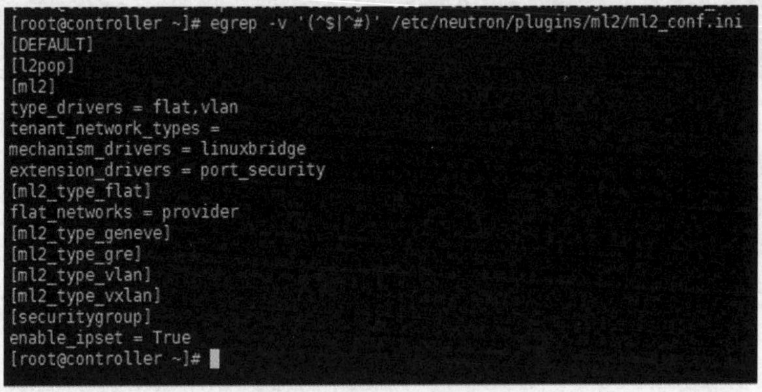

图 6-119　查看生效配置 2

(4) 快速配置 /etc/neutron/plugins/ml2/linuxbridge_agent.ini。

```
# openstack-config --set /etc/neutron/plugins/ml2/linuxbridge_agent.ini linux_bridge physical_interface_mappings provider:eth0
# openstack-config --set /etc/neutron/plugins/ml2/linuxbridge_agent.ini vxlan enable_vxlan False
# openstack-config --set /etc/neutron/plugins/ml2/linuxbridge_agent.ini securitygroup enable_security_group True
# openstack-config --set /etc/neutron/plugins/ml2/linuxbridge_agent.ini securitygroup firewall_driver neutron.agent.linux.iptables_firewall.IptablesFirewallDriver
```

虚拟机网卡名称要改为用户自己的虚拟机网卡名称，不然控制结点的 linuxbridge_agent 的服务无法启动，这里网卡名为 eth0，可以利用 ifconfig 命令查看，如图 6-120 所示。

图 6-120　ifconfig 命令查看

查看生效的配置，如图 6-121 所示。

[root@controller ~]# egrep -v '(^$|^#)' /etc/neutron/plugins/ml2/linuxbridge_agent.ini

图 6-121　查看生效配置 3

（5）快速配置 /etc/neutron/dhcp_agent.ini。

openstack-config --set /etc/neutron/dhcp_agent.ini DEFAULT interface_driver linuxbridge

openstack-config --set /etc/neutron/dhcp_agent.ini DEFAULT dhcp_driver neutron.agent.linux.dhcp.Dnsmasq

openstack-config --set /etc/neutron/dhcp_agent.ini DEFAULT enable_isolated_metadata True

查看生效的配置，如图 6-122 所示。

[root@controller ~]# egrep -v '(^$|^#)' /etc/neutron/dhcp_agent.ini

图 6-122　查看生效配置 4

（6）快速配置 /etc/neutron/metadata_agent.ini。

openstack-config --set /etc/neutron/metadata_agent.ini DEFAULT

nova_metadata_host controller
　　# openstack-config --set /etc/neutron/metadata_agent.ini DEFAULT metadata_proxy_shared_secret neutron

查看生效的配置，如图 6-123 所示。

[root@controller ~]# egrep -v '(^$|^#)' /etc/neutron/metadata_agent.ini

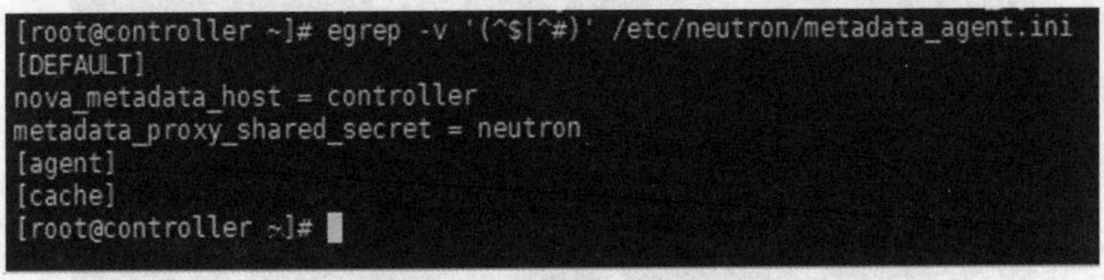

```
[root@controller ~]# egrep -v '(^$|^#)' /etc/neutron/metadata_agent.ini
[DEFAULT]
nova_metadata_host = controller
metadata_proxy_shared_secret = neutron
[agent]
[cache]
[root@controller ~]#
```

图 6-123　查看生效配置 5

（7）配置计算服务使用网络服务，将 neutron 添加到计算结点中。

　　# openstack-config --set /etc/nova/nova.conf neutron url http://controller:9696

　　# openstack-config --set /etc/nova/nova.conf neutron auth_url http://controller:5000

　　# openstack-config --set /etc/nova/nova.conf neutron auth_type password

　　# openstack-config --set /etc/nova/nova.conf neutron project_domain_name default

　　# openstack-config --set /etc/nova/nova.conf neutron user_domain_name default

　　# openstack-config --set /etc/nova/nova.conf neutron region_name RegionOne

　　# openstack-config --set /etc/nova/nova.conf neutron project_name service

　　# openstack-config --set /etc/nova/nova.conf neutron username neutron

　　# openstack-config --set /etc/nova/nova.conf neutron password neutron

　　# openstack-config --set /etc/nova/nova.conf neutron service_metadata_proxy true

　　# openstack-config --set /etc/nova/nova.conf neutron metadata_proxy_shared_secret neutron

查看生效的配置，如图 6-124 所示。

```
[root@controller ~]# egrep -v '(^$|^#)' /etc/nova/nova.conf
```

```
[root@controller ~]# egrep -v '(^$|^#)' /etc/nova/nova.conf
[DEFAULT]
enabled_apis = osapi_compute,metadata
my_ip = 10.0.0.9
use_neutron = true
firewall_driver = nova.virt.firewall.NoopFirewallDriver
transport_url = rabbit://openstack:openstack@controller
[api]
auth_strategy = keystone
[api_database]
connection = mysql+pymysql://nova:nova@controller/nova_api
[barbican]
[cache]
[cells]
[cinder]
[compute]
[conductor]
[console]
[consoleauth]
[cors]
[database]
connection = mysql+pymysql://nova:nova@controller/nova
[devices]
[ephemeral_storage_encryption]
[filter_scheduler]
[glance]
api_servers = http://controller:9292
[guestfs]
[healthcheck]
[hyperv]
[ironic]
[key_manager]
[keystone]
[keystone_authtoken]
auth_url = http://controller:5000/v3
memcached_servers = controller:11211
auth_type = password
project_domain_name = default
user_domain_name = default
project_name = service
username = nova
password = nova
[libvirt]
```

图 6-124　查看生效配置 6

（8）初始化安装网络插件。

创建网络插件的连接，初始化网络的脚本插件会用到 /etc/neutron/plugin.ini，需要使用 ML2 的插件进行提供。

```
[root@controller ~]# ln -s /etc/neutron/plugins/ml2/ml2_conf.ini /etc/neutron/plugin.ini
```

（9）同步数据库如图 6-125 所示。

```
su -s /bin/sh -c "neutron-db-manage --config-file /etc/neutron/neutron.conf \--config-file /etc/neutron/plugins/ml2/ml2_conf.ini upgrade head" neutron
```

图 6-125 同步数据库

(10) 重启 nova_api 服务。

[root@controller ~]# systemctl restart openstack-nova-api.service

(11) 启动 neutron 服务并设置开机启动，如图 6-126~图 6-128 所示。

systemctl start neutron-server.service neutron-linuxbridge-agent.service neutron-dhcp-agent.service neutron-metadata-agent.service

systemctl status neutron-server.service neutron-linuxbridge-agent.service neutron-dhcp-agent.service neutron-metadata-agent.service

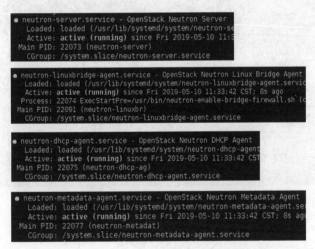

图 6-126 启动 neutron 服务并设置开机启动 1

systemctl enable neutron-server.service neutron-linuxbridge-agent.service neutron-dhcp-agent.service neutron-metadata-agent.service

图 6-127 启动 neutron 服务并设置开机启动 2

systemctl list-unit-files |grep neutron* |grep enabled

```
[root@controller ~]# systemctl list-unit-files |grep neutron* |grep enabled
neutron-dhcp-agent.service                    enabled
neutron-linuxbridge-agent.service             enabled
neutron-metadata-agent.service                enabled
neutron-server.service                        enabled
[root@controller ~]#
```

图 6-128　启动 neutron 服务并设置开机启动 3

6.4.3　安装和配置计算结点网络服务

（1）安装计算结点 Neutron 网络组件。

在计算结点执行以下命令完成安装 Neutron 服务软件包功能，如图 6-129 所示。

```
# yum install openstack-neutron-linuxbridge ebtables ipset -y
```

```
[root@compute ~]# yum install openstack-neutron-linuxbridge ebtables ipset -y
Loaded plugins: fastestmirror
Loading mirror speeds from cached hostfile
 * base: mirrors.aliyun.com
 * centos-qemu-ev: centos.ustc.edu.cn
 * extras: mirrors.aliyun.com
 * updates: mirrors.aliyun.com
Package ebtables-2.0.10-16.el7.x86_64 already installed and latest version
Resolving Dependencies
--> Running transaction check
---> Package ipset.x86_64 0:6.38-3.el7_6 will be installed
--> Processing Dependency: ipset-libs(x86-64) = 6.38-3.el7_6 for package: ipset-6.38-3.el7_6.x86_64
--> Processing Dependency: libipset.so.11(LIBIPSET_4.6)(64bit) for package: ipset-6.38-3.el7_6.x86_64
--> Processing Dependency: libipset.so.11(LIBIPSET_4.5)(64bit) for package: ipset-6.38-3.el7_6.x86_64
--> Processing Dependency: libipset.so.11(LIBIPSET_3.0)(64bit) for package: ipset-6.38-3.el7_6.x86_64
--> Processing Dependency: libipset.so.11(LIBIPSET_2.0)(64bit) for package: ipset-6.38-3.el7_6.x86_64
--> Processing Dependency: libipset.so.11(LIBIPSET_1.0)(64bit) for package: ipset-6.38-3.el7_6.x86_64
--> Processing Dependency: libipset.so.11()(64bit) for package: ipset-6.38-3.el7_6.x86_64
---> Package openstack-neutron-linuxbridge.noarch 1:13.0.3-1.el7 will be installed
--> Processing Dependency: openstack-neutron-common = 1:13.0.3-1.el7 for package: 1:openstack-neutron-
ch
--> Running transaction check
---> Package ipset-libs.x86_64 0:6.38-3.el7_6 will be installed
```

图 6-129　安装和配置计算结点网络服务

（2）快速配置 /etc/neutron/neutron.conf。

　　#openstack-config --set /etc/neutron/neutron.conf DEFAULT transport_url rabbit: //openstack: openstack@controller

　　# openstack-config --set /etc/neutron/neutron.conf DEFAULT auth strategy keystone

　　#openstack-config --set /etc/neutron/neutron.conf keystone_authtoken www_authenticate_uri http: //controller: 5000

　　#openstack-config --set /etc/neutron/neutron.conf keystone_authtoken auth_url http: //controller: 5000

　　#openstack-config --set /etc/neutron/neutron.conf keystone_authtoken memcached_servers controller: 11211

　　#openstack-config --set /etc/neutron/neutron.conf keystone_authtoken

auth_type password
   #openstack-config --set /etc/neutron/neutron.conf keystone_authtoken project_domain_name default
   #openstack-config --set /etc/neutron/neutron.conf keystone_authtoken user_domain_name default
   #openstack-config --set /etc/neutron/neutron.conf keystone_authtoken project_name service
   #openstack-config --set /etc/neutron/neutron.conf keystone_authtoken username neutron
   #openstack-config --set /etc/neutron/neutron.conf keystone_authtoken password neutron
   #openstack-config --set /etc/neutron/neutron.conf oslo_concurrency lock_path /var/lib/neutron/tmp

查看生效的配置，如图 6-130 所示。
[root@compute ~]# egrep -v '(^$|^#)' /etc/neutron/neutron.conf

```
[root@compute ~]# egrep -v '(^$|^#)' /etc/neutron/neutron.conf
[DEFAULT]
transport_url = rabbit://openstack:openstack@controller
auth_strategy = keystone
[agent]
[cors]
[database]
[keystone_authtoken]
www_authenticate_uri = http://controller:5000
auth_url = http://controller:5000
memcached_servers = controller:11211
auth_type = password
project_domain_name = default
user_domain_name = default
project_name = service
username = neutron
password = neutron
[matchmaker_redis]
[nova]
[oslo_concurrency]
lock_path = /var/lib/neutron/tmp
[oslo_messaging_amqp]
[oslo_messaging_kafka]
[oslo_messaging_notifications]
[oslo_messaging_rabbit]
[oslo_messaging_zmq]
[oslo_middleware]
[oslo_policy]
[quotas]
[ssl]
[root@compute ~]#
```

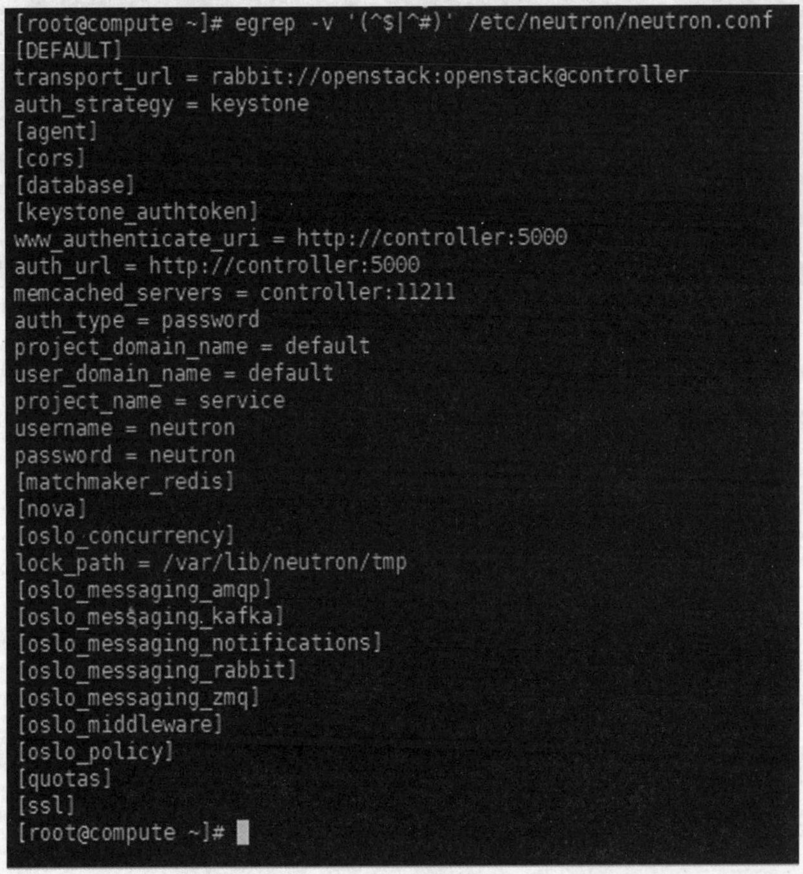

图 6-130　查看生效配置 1

(3) 快速配置 /etc/neutron/plugins/ml2/linuxbridge_agent.ini。

　　#openstack-config --set /etc/neutron/plugins/ml2/linuxbridge_agent.ini linux_bridge physical_interface_mappings provider：eth0

　　#openstack-config --set /etc/neutron/plugins/ml2/linuxbridge_agent.ini vxlan enable_vxlan false

　　#openstack-config --set /etc/neutron/plugins/ml2/linuxbridge_agent.ini securitygroup enable_security_group true

　　#openstack-config --set /etc/neutron/plugins/ml2/linuxbridge_agent.ini securitygroup firewall_driver neutron.agent.linux.iptables_firewall.IptablesFirewallDriver

第一个选项要配置计算结点自身的网卡名称，这里是eth0，可以使用ifconfig命令查看网卡名称。查看生效的配置，如图6-131所示。

　　egrep -v '(^$|^#)' /etc/neutron/plugins/ml2/linuxbridge_agent.ini

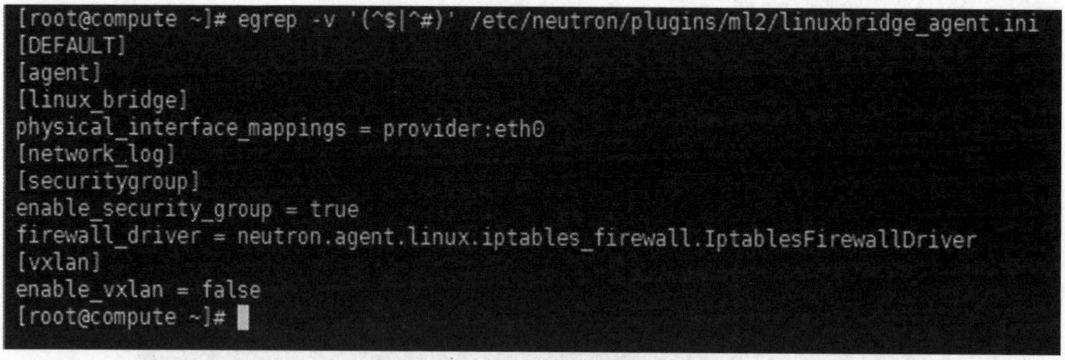

```
[root@compute ~]# egrep -v '(^$|^#)' /etc/neutron/plugins/ml2/linuxbridge_agent.ini
[DEFAULT]
[agent]
[linux_bridge]
physical_interface_mappings = provider:eth0
[network_log]
[securitygroup]
enable_security_group = true
firewall_driver = neutron.agent.linux.iptables_firewall.IptablesFirewallDriver
[vxlan]
enable_vxlan = false
[root@compute ~]#
```

图6-131　查看生效配置2

(4) 配置nova计算服务与neutron网络服务协同工作。

　　#openstack-config --set /etc/nova/nova.conf neutron url http://controller：9696

　　#openstack-config --set /etc/nova/nova.conf neutron auth_url http://controller：5000

　　#openstack-config --set /etc/nova/nova.conf neutron auth_type password

　　#openstack-config --set /etc/nova/nova.conf neutron project_domain_name default

　　#openstack-config --set /etc/nova/nova.conf neutron user_domain_name default

　　#openstack-config --set /etc/nova/nova.conf neutron region_name RegionOne

　　#openstack-config --set /etc/nova/nova.conf neutron project_name service

```
#openstack-config --set /etc/nova/nova.conf neutron username neutron
#openstack-config --set /etc/nova/nova.conf neutron password neutron
```

查看生效的配置，如图 6-132 所示。

```
[root@compute ~]# egrep -v '(^$|^#)' /etc/nova/nova.conf
```

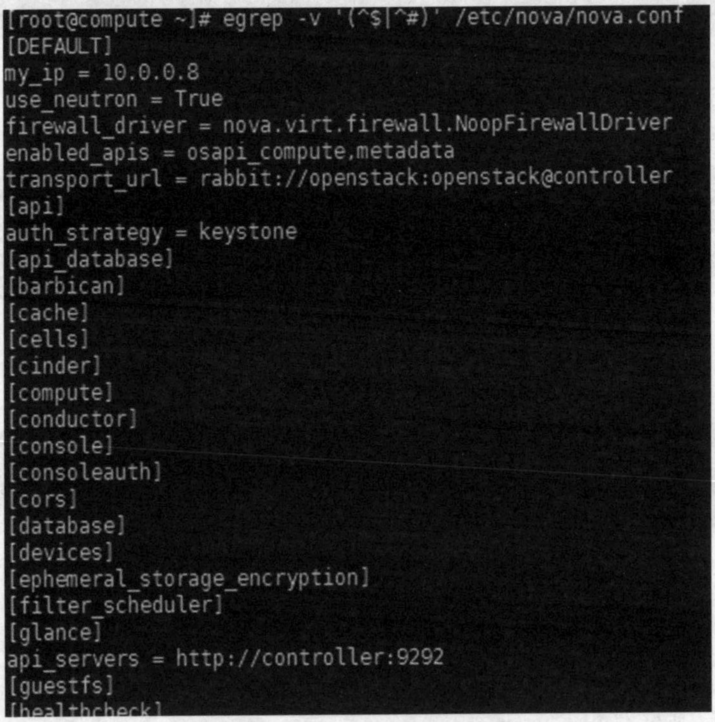

图 6-132　查看生效配置 3

(5) 重启计算结点，如图 6-133 所示。

```
systemctl restart openstack-nova-compute.service
systemctl status openstack-nova-compute.service
```

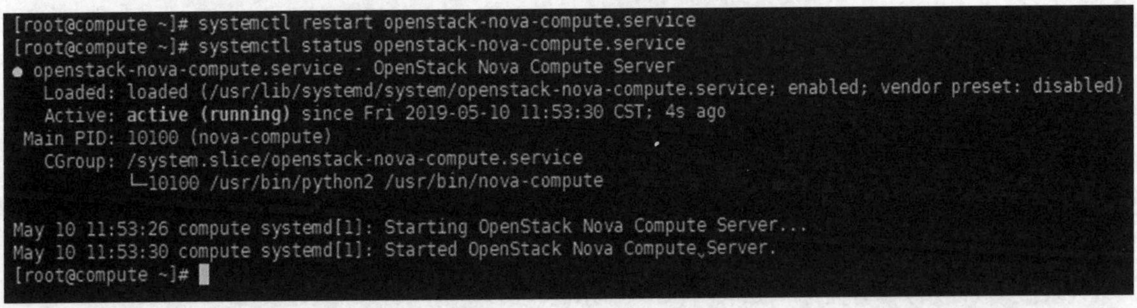

图 6-133　重启计算结点

(6) 启动 neutron 网络组件，并配置开机自启动，如图 6-134~图 6-136 所示。

```
#systemctl restart neutron-linuxbridge-agent.service
```

```
#systemctl status neutron-linuxbridge-agent.service
```

```
[root@compute ~]# systemctl restart neutron-linuxbridge-agent.service
[root@compute ~]# systemctl status neutron-linuxbridge-agent.service
● neutron-linuxbridge-agent.service - OpenStack Neutron Linux Bridge Agen
   Loaded: loaded (/usr/lib/systemd/system/neutron-linuxbridge-agent.serv
   Active: active (running) since Fri 2019-05-10 11:55:03 CST; 2s ago
  Process: 10148 ExecStartPre=/usr/bin/neutron-enable-bridge-firewall.sh
 Main PID: 10164 (neutron-linuxbr)
   CGroup: /system.slice/neutron-linuxbridge-agent.service
```

图 6-134　启动 neutron 网络组件 1

```
#systemctl enable neutron-linuxbridge-agent.service
```

```
[root@compute ~]# systemctl enable neutron-linuxbridge-agent.service
Created symlink from /etc/systemd/system/multi-user.target.wants/neutron-linuxbrid
on-linuxbridge-agent.service.
[root@compute ~]#
```

图 6-135　启动 neutron 网络组件 2

```
#systemctl list-unit-files |grep neutron* |grep enabled
```

```
[root@compute ~]# systemctl list-unit-files |grep neutron* |grep enabled
neutron-linuxbridge-agent.service             enabled
[root@compute ~]#
```

图 6-136　启动 neutron 网络组件 3

6.4.4　验证网络服务

验证网络服务的命令有：

列举 neutron 资源列表
```
# neutron ext-list
```

查询内部 dnsmasq 的进程信息
```
# ps -ef |grep dnsmasq
```

查看 linux 网桥信息
```
# brctl show
```

查看 Neutron 虚拟路由器信息
```
# neutron router-list
```

列举 neutron agent 列表
```
# neutron agent-list
```

在控制结点检查确认 neutron 服务安装启动，获取管理权限：
```
[root@controller ~]# source keystone-admin-pass.sh
```

列表查看加载的网络插件，如图 6-137 所示。
```
[root@controller ~]# openstack extension list --network
```

单元 6　安装 OpenStack 服务

```
[root@controller ~]# openstack extension list --network
+------------------------------------------------------------+
| Name                                                       |
+------------------------------------------------------------+
| Default Subnetpools                                        |
| Network IP Availability                                    |
| Network Availability Zone                                  |
| Network MTU (writable)                                     |
| Port Binding                                               |
| agent                                                      |
| Subnet Allocation                                          |
| DHCP Agent Scheduler                                       |
| Neutron external network                                   |
| Neutron Service Flavors                                    |
| Network MTU                                                |
| Availability Zone                                          |
| Quota management support                                   |
| Tag support for resources with standard attribute: subnet, tr|
| Availability Zone Filter Extension                         |
| If-Match constraints based on revision_number              |
| Filter parameters validation                               |
| Multi Provider Network                                     |
```

图 6-137　查看加载的网络插件

可以使用另一个命令查看简略信息，如图 6-138 所示。

```
[root@controller ~]# neutron ext-list
```

```
[root@controller ~]# neutron ext-list
neutron CLI is deprecated and will be removed in the future. Use openstack CLI instead.
+---------------------------+-----------------------------+
| alias                     | name                        |
+---------------------------+-----------------------------+
| default-subnetpools       | Default Subnetpools         |
| network-ip-availability   | Network IP Availability     |
| network_availability_zone | Network Availability Zone   |
| net-mtu-writable          | Network MTU (writable)      |
| binding                   | Port Binding                |
| agent                     | agent                       |
| subnet_allocation         | Subnet Allocation           |
| dhcp_agent_scheduler      | DHCP Agent Scheduler        |
| external-net              | Neutron external network    |
| flavors                   | Neutron Service Flavors     |
| net-mtu                   | Network MTU                 |
| availability_zone         | Availability Zone           |
| quotas                    | Quota management support    |
| standard-attr-tag         | Tag support for resources with standard attribute: s|
tpool, port, security_group, floatingip
```

图 6-138　查看简略信息

查看网络代理列表，如图 6-139 所示。

```
[root@controller ~]# openstack network agent list
```

```
[root@controller ~]# openstack network agent list
+--------------------------------------+--------------------+------------+-------------------+-------+-------+-------------------------
------+
| ID                                   | Agent Type         | Host       | Availability Zone | Alive | State | Binary
      |
+--------------------------------------+--------------------+------------+-------------------+-------+-------+-------------------------
------+
| 5c7366d0-3895-4459-90ca-85b12a872a96 | Metadata agent     | controller | None              | :-)   | UP    | neutron-metadata-age
nt    |
| 8dcdbc53-58a6-444c-9d07-6b6612fb3db7 | DHCP agent         | controller | nova              | :-)   | UP    | neutron-dhcp-agent
      |
| c855d332-e67d-4469-8ac4-e0c62aa7a916 | Linux bridge agent | compute    | None              | :-)   | UP    | neutron-linuxbridge-
agent |
| fef30d8c-fe59-4d8c-940b-b04358dd2f04 | Linux bridge agent | controller | None              | :-)   | UP    | neutron-linuxbridge-
agent |
+--------------------------------------+--------------------+------------+-------------------+-------+-------+-------------------------
------+
[root@controller ~]#
```

图 6-139　查看网络代理列表

正常情况下控制结点有三个服务，计算结点有一个服务。如果有差异，则需要检查计算结点网卡名称、ip 地址、端口等。

6.5 Dashboard 的安装配置

Dashboard 是一个 web 接口，又称为 Horizon，使得云平台管理员以及用户可以管理不同的 OpenStack 资源以及服务。这个部署示例使用的是 Apache Web 服务器。Dashboard 仅在核心服务中要求认证服务。用户可以将 Dashboard 与其他服务，如镜像服务、计算服务和网络服务等结合使用，也可以在单机服务环境如对象存储中使用 Dashboard。

除了 Dashboard，用户也可以直接使用 OpenStack 命令行客户端管理和使用 OpenStack。这里将在控制结点安装 Dashboard。

Dashboard 特点如下：

（1）提供一个 Web 界面操作 OpenStack 的系统。

（2）使用 Django 框架基于 OpenStack API 开发。

（3）支持将 Session 存储在 DB、Memcached。

（4）支持集群。

6.5.1 Dashboard 安装和配置组件

1. 安装 Dashboard

运行以下命令安装 Dashboard：

```
[root@controller ~]# yum install -y  openstack-dashboard
```

2. 配置 Dashboard

（1）编辑 Dashboard 配置文件。

```
vi /etc/openstack-dashboard/local_settings
# 允许所有主机访问 Dashboard:
```

```
ALLOWED_HOSTS=['*', ]
# session 引擎
SESSION_ENGINE='django.contrib.sessions.backends.cache'
# 配置 API 版本
OPENSTACK_API_VERSIONS={
    "identity": 3,
    "image": 2,
    "volume": 2,
}
# 指定控制结点的主机名：
OPENSTACK_HOST="controller"
# 使用第三版 Keystone API
OPENSTACK_KEYSTONE_URL="http://%s:5000/v3" % OPENSTACK_HOST
# 仪表盘创建用户默认配置为 user
OPENSTACK_KEYSTONE_DEFAULT_ROLE="user"
# 启用对域的支持
OPENSTACK_KEYSTONE_MULTIDOMAIN_SUPPORT=True
# 通过仪表盘创建用户的默认域配置为 default
OPENSTACK_KEYSTONE_DEFAULT_DOMAIN="default"
# 配置 memecached 存储服务
CACHES={
    'default': {
        'BACKEND': 'django.core.cache.backends.memcached.MemcachedCache',
        'LOCATION': 'controller:11211',
    }
}
# 配置网络（需要添加几项）
OPENSTACK_NEUTRON_NETWORK={
    'enable_router': False,
    'enable_quotas': False,
    'enable_distributed_router': False,
    'enable_ha_router': False,
    'enable_fip_topology_check': False,
    'enable_lb': False,
    'enable_firewall': False,
    'enable_vpn': False,
}
```

\# 将时区修改为亚洲上海：

TIME_ZONE="Asia/Shanghai"

（2）修改 /etc/httpd/conf.d/openstack-dashboard.conf。

vi /etc/httpd/conf.d/openstack-dashboard.conf

添加配置如下：

WSGIApplicationGroup %{GLOBAL}

6.5.2　验证 Dashboard

启动 Web 服务器和 Memcached 服务：

[root@controller ~]# systemctl restart httpd.service memcached.service

[root@controller ~]# systemctl status httpd.service memcached.service

访问 Dashboard 需打开浏览器并输入 http：//10.0.0.9/dashboard，用户名处输入 admin 或 myuser，这两个用户在配置认证服务时创建，其中 admin 为管理用户，myuser 为普通用户，密码都是 123456，域名是 default，如图 6-140 所示。

图 6-140　登录界面

6.6　块存储服务 Cinder 的安装配置

Cinder 是在虚拟机和具体存储设备之间引入一层"逻辑存储卷"的抽象，Cinder 本身并不是一种存储技术，只是提供一个中间的抽象层，Cinder 通过调用不同存储后端类型的驱动接口来管理相对应的后端存储，为用户提供统一的卷相关操作的存储接口。Cinder 存储管理架构如图 6-141 所示。

单元 6　安装 OpenStack 服务

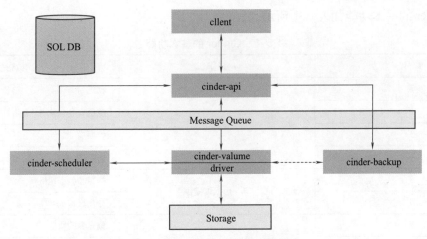

图 6-141　Cinder 存储管理架构

（1）为虚拟机实例提供 volume 卷的块存储服务：

一个 volume 可以同时挂载到多个实例上，作为虚拟机实例的本地磁盘来使用；

共享的卷同时只能被一个实例进行写操作；

它是一个资源管理系统，负责向虚拟机提供持久块存储资源；

它把不同的后端存储进行封装，向外提供统一的 API。

它不是新开发的块设备存储系统，而是使用插件的方式，结合不同后端存储的驱动提供块存储服务。主要核心是对卷的管理，允许对卷、卷的类型、卷的快照进行处理。Cinder 的内部架构如图 6-142 所示。

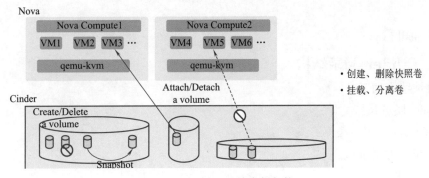

图 6-142　Cinder 的内部架构

（2）三个主要组成部分：

① cinder-api 组件负责向外提供 Cinder REST API。

② cinder-scheduler 组件负责分配存储资源。

③ cinder-volume 组件负责封装 driver，不同的 driver 负责控制不同的后端存储。

（3）组件之间的 RPC 靠消息队列（Queue）实现。

Cinder 的开发工作主要集中在 scheduler 和 driver，以便提供更多的调度算法、更多的功能，以及指出更多的后端存储。

（4）Volume 元数据和状态保存在 Database 中。

(5) Cinder 的基本功能如表 6-1 所示。

表 6-1　Cinder 的基本功能

序号	操作	功能
1	卷操作	创建卷
2		从已有卷创建卷（克隆）
3		扩展卷
4		删除卷
5	卷-虚拟机操作	挂载卷到虚拟机
6		分离虚拟机卷
7	卷-快照操作	创建卷的快照
8		从已有卷快照创建卷
9		删除快照
10	卷-镜像操作	从镜像创建卷
11		从卷创建镜像

(6) 支持的文件系统类型：

- LVM / ISCSI；
- NFS；
- NetAPP NFS；
- Gluster；
- DELL Equall Logic。

6.6.1　Cinder 目录结构

Cinder 目录结构如下：

Bin：Cinder 服务执行文件。

Cinder：Cinder 代码目录。

Contrib：第三方贡献包。

Doc：技术文档。

Etc：Cinder 配置文件样例。

Tools：工具。

Babel.cfg：Flask-Babel 配置。

CONTRIBUTING.md：贡献指南。

HACKING.rst：Hack 指南。

LICENSE：Apache2 LICENSE。

MANIFEST.in：打包规则。

OpenStack-common.conf：OSLO 配置。

Pylintre：Pylint 代码分析配置。
README.rst：Cinder 简介。
Runjests.sh：测试案例。
Setup.cfg：setup.py 配置。
Setup.py：Cinder 安装脚本。
Tox.ini：Python 的标准化测试。

6.6.2 安装和配置控制结点

在安装和配置块存储服务之前，用户必须创建数据库、服务证书和 API 端点。

1. 完成下面的步骤以创建数据库

(1) 用数据库连接客户端以 root 用户连接到数据库服务器。

```
# mysql -u root -p123456
```

(2) 创建 cinder 数据库。

```
CREATE DATABASE cinder;
```

(3) 允许 cinder 数据库合适的访问权限。

```
GRANT ALL PRIVILEGES ON cinder.* TO 'cinder'@'localhost' IDENTIFIED BY 'cinder';
GRANT ALL PRIVILEGES ON cinder.* TO 'cinder'@'%' IDENTIFIED BY 'cinder';
flush privileges;
```

退出数据库客户端。

(4) 获得 admin 凭证来获取只有管理员能执行的命令的访问权限。

```
# source keystone-admin-pass.sh
```

2. 完成以下步骤创建服务证书

(1) 创建一个 cinder 用户，如图 6-143 所示。

```
# openstack user create --domain default --password=cinder cinder
```

图 6-143 创建 Cinder 用户

(2) 添加 admin 角色到 cinder 用户上。

```
$ openstack role add --project service --user cinder admin
```

(3) 创建 cinder 和 cinderv2 服务实体，如图 6-144 所示。

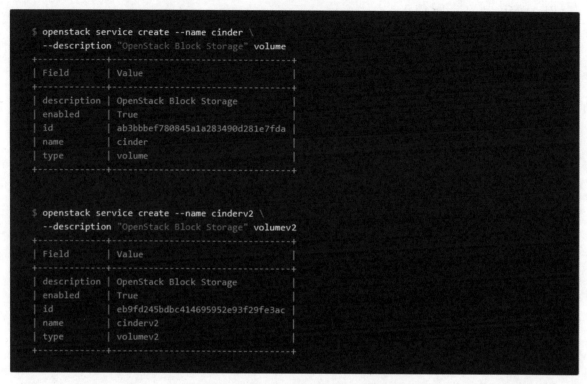

图 6-144　创建 cinder 和 cinder 2 服务实体

(4) 创建 cinder 服务的 API 端点（Endpoint）。

openstack endpoint create --region RegionOne volumev public http://controller:8776/v2/%\(project_id\)s

openstack endpoint create --region RegionOne volumev internal http://controller:8776/v2/%\(project_id\)s

openstack endpoint create --region RegionOne volumev admin http://controller:8776/v2/%\(project_id\)s

openstack endpoint create --region RegionOne volumev2 public http://controller:8776/v3/%\(project_id\)s

openstack endpoint create --region RegionOne volumev2 internal http://controller:8776/v3/%\(project_id\)s

openstack endpoint create --region RegionOne volumev2 admin http://controller:8776/v3/%\(project_id\)s

openstack endpoint list

3. 安装并配置 cinder 相关组件

(1) 安装软件包。

yum install openstack-cinder -y

(2) 快速修改 cinder 配置。

```
# openstack-config --set /etc/cinder/cinder.conf database connection mysql+pymysql://cinder:cinder@controller/cinder
# openstack-config --set /etc/cinder/cinder.conf DEFAULT transport_url rabbit://openstack:openstack@controller
# openstack-config --set /etc/cinder/cinder.conf DEFAULT auth_strategy keystone
# openstack-config --set /etc/cinder/cinder.conf keystone_authtoken auth_uri http://controller:5000
# openstack-config --set /etc/cinder/cinder.conf keystone_authtoken auth_url http://controller:5000
# openstack-config --set /etc/cinder/cinder.conf keystone_authtoken memcached_servers controller:11211
# openstack-config --set /etc/cinder/cinder.conf keystone_authtoken auth_type password
# openstack-config --set /etc/cinder/cinder.conf keystone_authtoken project_domain_name default
# openstack-config --set /etc/cinder/cinder.conf keystone_authtoken user_domain_name default
# openstack-config --set /etc/cinder/cinder.conf keystone_authtoken project_name service
# openstack-config --set /etc/cinder/cinder.conf keystone_authtoken username cinder
# openstack-config --set /etc/cinder/cinder.conf keystone_authtoken password cinder
```

(3) 配置控制结点 ip。

```
# openstack-config --set /etc/cinder/cinder.conf DEFAULT my_ip 10.0.0.9
# openstack-config --set /etc/cinder/cinder.conf oslo_concurrency lock_path /var/lib/nova/tmp
```

(4) 检查生效的配置。

```
# grep '^[a-z]' /etc/cinder/cinder.conf
```

(5) 初始化块设备服务的数据库。

```
# su -s /bin/sh -c "cinder-manage db sync" cinder
```

(6) 修改 nova 配置文件，配置 nova 调用 cinder 服务。

```
openstack-config --set /etc/nova/nova.conf cinder os_region_name RegionOne
```

(7) 重启计算 API 服务。

```
# systemctl restart openstack-nova-api.service
```

(8) 启动块设备存储服务，并将其配置为开机自启。

```
# systemctl enable openstack-cinder-api.service openstack-cinder-scheduler.service
# systemctl start openstack-cinder-api.service openstack-cinder-scheduler.service
# systemctl status openstack-cinder-api.service openstack-cinder-scheduler.service
```

6.6.3 安装和配置存储结点

在安装和配置块存储服务之前，用户必须准备好存储设备。存储结点建议单独部署服务器（最好是物理机），测试时也可以部署在控制结点或者计算结点。本书中，存储结点使用 LVM 逻辑卷提供服务，需要提供一块空的磁盘用以创建 LVM 逻辑卷。

1. 安装支持的工具包

(1) 安装 LVM 相关软件包。

```
# yum install lvm2 device-mapper-persistent-data -y
```

(2) 启动 LVM 的 metadata 服务并设置该服务随系统启动。

```
# systemctl start lvm2-lvmetad.service
# systemctl enable lvm2-lvmetad.service
```

(3) 创建 LVM 物理卷 /dev/sdb。

```
# pvcreate /dev/sdb
Physical volume "/dev/sdb" successfully created
```

(4) 创建 LVM 卷组 cinder-volumes，块存储服务会在这个卷组中创建逻辑卷。

```
# vgcreate cinder-volumes /dev/sdb
Volume group "cinder-volumes" successfully created
```

只有 OpenStack 实例可以访问块存储卷组。不过，底层的操作系统管理这些设备并将其与卷关联。默认情况下，LVM 卷扫描工具会扫描"/dev"目录，查找包含 LVM 卷的块存储设备。如果其他项目在它们的卷上使用 LVM，扫描工具检测到这些卷时会尝试缓存它们，可能会在底层操作系统和项目卷上产生各种问题。用户必须重新配置 LVM，让它只扫描包含"cinder-volume"卷组的设备。编辑"/etc/lvm/lvm.conf"文件并完成下面的操作。在"devices"部分，添加一个过滤器，只接受"/dev/sdb"设备，拒绝其他所有设备：

```
vim /etc/lvm/lvm.conf
devices {...filter=[ "a/sdb/", "r/.*/"]
```

配置规则如下：

每个过滤器组中的元素都以"a"开头，即为 accept，或以 r 开头，即为 **reject**，并且包括一个设备名称的正则表达式规则。过滤器组必须以"r/.*/"结束，过滤所有保留设备。用户可以使用命令：'vgs -vvvv'来测试过滤器。

【注意】

如果用户的存储结点在操作系统磁盘上使用了 LVM，还必须添加相关的设备到过滤器中。例如，/dev/sda 设备包含操作系统命令：

```
filter=[ "a/sda/", "a/sdb/", "r/.*/"]
```
类似的，如果用户的计算结点在操作系统磁盘上使用了 LVM 卷组，则必须修改这些结点上 /etc/lvm/lvm.conf 文件中的过滤器，将操作系统磁盘包含到过滤器中。例如："/dev/sda"设备包含操作系统命令：

```
filter=[ "a/sda/", "r/.*/"]
```

2. 在存储结点安装并配置 cinder 组件

(1) 安装软件包。

```
# yum install openstack-cinder targetcli python-keystone -y
```

(2) 在存储结点快速修改 cinder 配置。

```
# openstack-config --set /etc/cinder/cinder.conf database connection mysql+pymysql://cinder:cinder@controller/cinder
# openstack-config --set /etc/cinder/cinder.conf DEFAULT transport_url rabbit://openstack:openstack@controller
# openstack-config --set /etc/cinder/cinder.conf DEFAULT auth_strategy keystone
# openstack-config --set /etc/cinder/cinder.conf keystone_authtoken www_authenticate_uri http://controller:5000
# openstack-config --set /etc/cinder/cinder.conf keystone_authtoken auth_url http://controller:5000
# openstack-config --set /etc/cinder/cinder.conf keystone_authtoken memcached_servers controller:11211
# openstack-config --set /etc/cinder/cinder.conf keystone_authtoken auth_type password
# openstack-config --set /etc/cinder/cinder.conf keystone_authtoken project_domain_name default
# openstack-config --set /etc/cinder/cinder.conf keystone_authtoken user_domain_name default
# openstack-config --set /etc/cinder/cinder.conf keystone_authtoken project_name service
# openstack-config --set /etc/cinder/cinder.conf keystone_authtoken username cinder
# openstack-config --set /etc/cinder/cinder.conf keystone_authtoken password cinder
# openstack-config --set /etc/cinder/cinder.conf DEFAULT my_ip 10.0.0.8
# openstack-config --set /etc/cinder/cinder.conf lvm volume_driver cinder.volume.drivers.lvm.LVMVolumeDriver
# openstack-config --set /etc/cinder/cinder.conf lvm volume_group
```

cinder-volumes

   # openstack-config --set /etc/cinder/cinder.conf lvm iscsi_protocol iscsi

   # openstack-config --set /etc/cinder/cinder.conf lvm iscsi_helper lioadm

   # openstack-config --set /etc/cinder/cinder.conf DEFAULT enabled_backends lvm

   # openstack-config --set /etc/cinder/cinder.conf DEFAULT glance_api_servers http://controller:9292

   # openstack-config --set /etc/cinder/cinder.conf oslo_concurrency lock_path /var/lib/cinder/tmp

【注意】

如果存储结点是双网卡，则选项 my_ip 需要配置存储结点的管理 IP，否则配置本机 IP。

3. 完成安装

启动块存储卷服务及其依赖的服务，并将其配置为随系统启动（两个服务）。

   # systemctl enable openstack-cinder-volume.service target.service

   # systemctl start openstack-cinder-volume.service target.service

   # systemctl status openstack-cinder-volume.service target.service

6.6.4 验证 Cinder 操作

这项操作在控制结点上执行，获得 admin 凭证来获取只有管理员能执行的命令的访问权限。

   # source keystone-admin-pass.sh

查看存储卷列表，如图 6-145 所示。

   # openstack volume service list

```
[root@controller ~]# openstack volume service list
+------------------+-------------+------+---------+-------+----------------------------+
| Binary           | Host        | Zone | Status  | State | Updated At                 |
+------------------+-------------+------+---------+-------+----------------------------+
| cinder-scheduler | controller  | nova | enabled | up    | 2019-05-13T20:08:42.000000 |
| cinder-volume    | block1@lvm  | nova | enabled | down  | 2019-05-13T08:19:54.000000 |
+------------------+-------------+------+---------+-------+----------------------------+
```

图 6-145 查看存储卷列表

单元 7

OpenStack 日常运维

学习目标

◎ 掌握 OpenStack 安装完成后的日常运维；
◎ 能进行 OpenStack 维护与诊断，包括控制结点、计算结点、网络诊断等。

7.1 控制结点的维护与排错

在前面章节中，学会了实战安装 OpenStack，若遇到 OpenStack 计划内或计划外的停机，影响是显而易见的。

对于日常运维来说，比较彻底的维护就是重启。控制结点重启后，可以通过以下命令确认所需的服务是否都已经正常运行。

```
[root@wz~]# ps aux | grep nova -
[root@wz~]# ps aux | grep glance -
[root@wz~]# ps aux | grep keystone
[root@wz~]# ps aux | grep cinder
```

确认服务是否正常工作的命令如下：

```
[root@wz~]# source openrc
[root@wz~]# glance index
[root@wz~]# nova list
[root@wz~]# keystone tenant - list
```

7.2 计算结点的维护与排错

计算结点重启后，可以通过以下命令确认所需的服务是否都已经正常运行。

```
[root@wz~]# ps aux | grep nova -compute
[root@wz~]# status nova-compute
```

7.3 网络诊断

7.3.1 检查网卡状态

在各个计算结点上，无论网卡是否被启用，都可以使用 ip a 命令来查看结点网卡的信息，如 IP 和 Vlan 的分配。在遇到一些网络问题时，检查网卡的信息状况都是必须要做的事情，相关实例命令如下：

```
[root@wz ~]# ip a | grep state
1: lo: <LOOPBACK,UP,LOWER_UP> mtu 65536 qdisc noqueue state UNKNOWN qlen 1
2: eno1: <BROADCAST,MULTICAST,UP,LOWER_UP> mtu 1500 qdisc mq portid 0894ef6b3694 state UP qlen 1000
3: eno2: <BROADCAST,MULTICAST,UP,LOWER_UP> mtu 1500 qdisc mq master br-wz portid 0894ef6b3695 state UP qlen 1000
4: br-wz: <BROADCAST,MULTICAST,UP,LOWER_UP> mtu 1500 qdisc noqueue state UP qlen 1000
5: br-849207184a8c: <NO-CARRIER,BROADCAST,MULTICAST,UP> mtu 1500 qdisc noqueue state DOWN
```

上述命令中，可以忽略 br-849207184a8c，它只是由 QEMU 创建的一个默认网桥，OpenStack 不使用。

7.3.2 虚拟机网络

登录虚拟机，使用 ping 命令测试外网，如输入 ping whwkzc.com。

```
[root@centos-7 ~]# ping whwkzc.com
PING whwkzc.com (58.49.94.97) 56(84) bytes of data.
64 bytes from 58.49.94.97 (58.49.94.97): icmp_seq=1 ttl=233 time=3.08 ms
64 bytes from 58.49.94.97 (58.49.94.97): icmp_seq=2 ttl=233 time=2.56 ms
64 bytes from 58.49.94.97 (58.49.94.97): icmp_seq=3 ttl=233 time=2.80 ms
```

```
64 bytes from 58.49.94.97 (58.49.94.97):icmp_seq=4 ttl=233 time=2.47 ms
64 bytes from 58.49.94.97 (58.49.94.97):icmp_seq=5 ttl=233 time=3.02 ms
64 bytes from 58.49.94.97 (58.49.94.97):icmp_seq=6 ttl=233 time=3.83 ms
64 bytes from 58.49.94.97 (58.49.94.97):icmp_seq=7 ttl=233 time=2.66 ms
```

Ping 数据包的路径如图 7-1 所示。

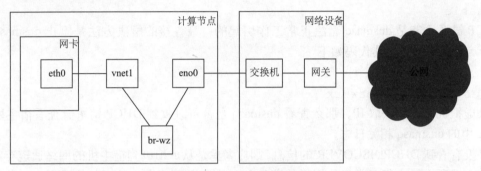

图 7-1　ping 数据包的路径

虚拟机发出 ping 命令后，虚拟机会产生网络数据包，并且在虚拟机内部虚拟出来的网卡中，如 eth0。数据包会被传递到宿主机的虚拟网卡，如 vnet1。然后，数据包会被送到宿主机中的虚拟网桥，如 br-wz 中。数据包会进一步发送到宿主机的物理网卡上。当数据包从宿主机发送到网关以后，OpenStack 即不再控制数据包的流向。如图 7-1 所示，网关为外部网关。上述流程的逆向过程即为 ping 命令的回复路径，每一个数据包都经过 4 个虚拟或物理的网络设备，其中任何一个设备出现故障，都会导致整个 OpenStack 网络故障。

7.3.3　检测网络

使用 ping 命令是快速检测网络是否正常的方法，如果能成功 ping 通外部网络，基本可以确认网络是没有故障的。如果不能 ping 通外部网络，排查方法为 ping 虚拟机的宿主机的 IP 地址。如果可以 ping 通，说明故障点可能出现在计算结点和网关之间，相反如果宿主机的 IP 都无法 ping 通，那么故障可能出现在虚拟机和宿主机之间，如连接虚拟网卡和物理网卡的网桥。尝试在本机的两个虚拟机之间互 ping，如果可以成功，则可能是防火墙的设置问题。

7.4　诊断 DHCP 和 DNS 故障

虚拟机启动后无法正常通过 dnsmasq 获取 IP 的故障排除，此命令是由 Nova 的网络服务启动。最直接的方法是查看虚拟机的控制台输出，其代码如下：

```
nova console-log <instance name or uuid>
```

如果虚拟机无法从 DHCP 中获取 IP，那么控制台输出中通常会有提示信息。例如：

```
udhcpc(v1.17.2)started
Sending discover...
Sending discover...
Sending discover...
No lease, forking to background
...
cloud-setup:checking http://192.168.0.191/2019-04-04/meta-data/instance-id
wget:can't connect to remote host(192.168.0.191):Network is unreachable
```

DHCP 错误可能是 dnsmasq 无法正常工作引起的，最直接的解决方法是停止 dnsmasq 进程，然后重新启动网络服务，其代码如下：

```
# killall dnsmasq
# restart nova-network
```

如果虚拟机仍无法获取 IP，那么查看 dnsmasq 是否可以收到 DHCP 请求。查看宿主机上 /var/log/syslog 中的 dnsmasq 相关日志。

如果没有看到 DHCPDISCOVER 的信息，则应考虑是从虚拟机到宿主机的网络通道存在问题。

如果上述所有日志都正常但依旧无法获取 IP，则问题可能出在数据包的返程上。若提示"no address available"，则说明 DHCP 服务已经没有剩余的 IP 可供分配了。

如果看不到 dnsmasq 的日志输出，则需要查看 dnsmasq 信息本身的状态，命令如下：

```
# ps aux | grep dnsmasq
```

如果 dnsmasq 自身没有任何问题，那么可以使用 tcpdump 逐一诊断网络查找是哪一部分丢失了 DHCP 请求。DHCP 是 UDP 请求，从客户端的 68 端口发送到 DHCP 服务端的 67 端口。开启虚拟机，然后监听所有虚拟网卡和物理网卡，查看哪个没有收到该 UDP 包，其代码如下：

```
# tcpdump -i br-wz -n port 67 or port 68
```

在 OpenStack 中，dnsmasq 充当着 DNS 服务器的角色。Dnsmasq 无法正常工作就会导致 DNS 服务无法工作。出现故障时，可以采用上述方法验证和修复 dnsmasq 服务。

7.4.1 日志与监控：故障的定位

OpenStack 集合了各种各样的系统组件，各组件之间相互协调调用。如当用户上传一个镜像时就需要同时有 nova-api、glance-api、Keystone 等之间的交互。这样，当出现故障时，很难定位具体是哪个组件出现故障，可参考以下方法进行排查。

首先可以排查的点就是与调用直接相关的日志。例如，当"nova list"命令调用失败，即查看最新的若干条 Nova 日志，具体代码参考如下：

```
# nova list
# tail -f /var/log/nova/nova-api.log
```

当发现故障与其他组件相关时，即可继续查看其他相关组件的日志。当日志显示错误提示 Nova 无法访问 Glance，即应该查看 glance_api 的日志，具体代码参考如下：

```
# nova list
```

```
# tail –f /var/log/glance/glance-api.log
```
按照此方法，依此类推，直到找出产生故障的组件。

7.4.2　错误日志

日志对于一个稳定的系统来说相当重要，对于 OpenStack 这样一个大型的系统，日志当然也是必不可少，理解 OpenStack 系统日志对于保证 OpenStack 环境稳定的重要性。对于出现系统错误，查看日志是一个很好的习惯。

OpenStack 通过生成大量日志信息来帮助排查系统安装运行期间出现的问题，接下来介绍几个常见服务的相关日志位置。在大部分 Linux 系统版本中，大部分日志文件都存放在 /var/log 子目录中，在 OpenStack 的控制结点中，各个组件的日志位置如下所示：

1. Nova 日志

OpenStack 计算服务日志位于 /var/log/nova，默认权限拥有者是 nova 用户。需要注意的是：并不是每台服务器上都包含所有的日志文件，如 nova-compute.log 仅在计算结点生成。

(1) nova-compute.log：虚拟机实例在启动和运行中产生的日志。

(2) nova-network.log：关于网络状态、分配、路由和安全组的日志。

(3) nova-manage.log：运行 nova-manage 命令时产生的日志。

(4) nova-scheduler.log：有关调度的，分配任务给结点以及消息队列的相关日志。

(5) nova-objectstore.log：镜像相关的日志。

(6) nova-api.log：用户与 OpenStack 交互以及 OpenStack 组件间交互的消息相关日志。

(7) nova-cert.log：nova-cert 过程的相关日志。

(8) nova-console.log：关于 nova-console 的 VNC 服务的详细信息。

(9) nova-consoleauth.log：关于 nova-console 服务的验证细节。

(10) nova-dhcpbridge.log：与 dhckbridge 服务先关的网络信息。

2. Dashboard 日志

Dashboard 是一个 DJango 的 Web 应用程序，默认运行在 Apache 服务器上，相应的运行日志也都记录在 Apache 的日志中，用户可以在 /var/log/apache2/ 中查看。

3. 存储日志

对象存储 Swift 默认日志写到 syslog 中，在 Ubuntu 系统中，可以通过 /var/log/syslog 查看，在其他系统中，可能位于 /var/log/messages 中。

块存储 Cinder 产生的日志默认存放在 /var/log/cinder 目录中。

(1) cinder-api.log：关于 cinder-api 服务的细节。

(2) cinder-scheduler.log：关于 cinder 调度服务的操作的细节。

(3) cinder-volume.log：与 cinder 卷服务相关的日志项。

4. Keystone 日志

身份认证 Keystone 服务的日志记录在 /var/log/keystone/keystone.log 中。

5. Glance 日志

镜像服务 Glance 的日志默认存放在 /var/log/glance 目录中。

(1) api.log：Glance API 相关的日志。

(2) registry.log：Glance registry 服务相关的日志。

根据日志配置的不同,会保存诸如元信息更新和访问记录这些信息。

6. Neutron 日志

网络服务 Neutron 的日志默认存放在 /var/log/neutron 目录中。

(1) dhcp-agent.log:关于 dhcp-agent 的日志。

(2) l3-agent.log:与 l3 代理及其功能相关的日志。

(3) metadata-agent.log:通过 neutron 代理给 Nova 元数据服务的相关日志。

(4) openvswitch-agent.log:与 openvswitch 相关操作的日志项,在具体实现 OpenStack 网络时,如果使用了不同的插件,就会有相应的日志文件名。

(5) server.log:与 Neutron API 服务相关的日志。

OpenStack 的日志格式都是统一的,如<时间戳><日志等级><代码模块><Request ID><日志内容><源代码位置>等。格式如下:

时间戳:日志记录的时间,包括年、月、日、时、分、秒、毫秒。

日志等级:有 INFO WARNING ERROR DEBUG 等。

代码模块:当前运行的模块。

Request ID:日志会记录连续不同的操作,为了便于区分和增加可读性,每个操作都被分配唯一的 Request ID,便于查找。

日志内容:这是日志的主体,记录当前正在执行的操作和结果等重要信息。

源代码位置:日志代码的位置,包括方法名称、源代码文件的目录位置和行号。这一项不是所有日志都有。

举例说明如下:

```
2018-12-10 20:46:49.566 DEBUG nova.virt.libvirt.config [req-5c973fff-e9ba-4317-bfd9-76678cc96584 None None] Generated XML ('<cpu>\n  <arch>x86_64</arch>\n  <model>Westmere</model>\n  <vendor>Intel</vendor>\n  <topology sockets="2" cores="3" threads="1"/>\n  <feature name="avx"/>\n  <feature name="ds"/>\n  <feature name="ht"/>\n  <feature name="hypervisor"/>\n  <feature name="osxsave"/>\n  <feature name="pclmuldq"/>\n  <feature name="rdtscp"/>\n  <feature name="ss"/>\n  <feature name="vme"/>\n  <feature name="xsave"/>\n</cpu>\n',) to_xml /opt/stack/nova/nova/virt/libvirt/config.py:82
```

由这条日志可得:代码模块是 nova.virt.libvirt.config,由此可知应该是 Hypervisor Libvirt 相关的操作。日志内容是生成 XML,如果要跟踪源代码,可以到 /opt/stack/nova/nova/virt/libvirt/config.py 的 82 行,方法是 to_xml:

```
def to_xml(self.pretty_print=ture):
    root=self.format_dom()
    xml_str=etree.tostring(root,pretty_print=pretty_print)
    LOG.debug("Generated XML %s",(xml_str,))
    return xml_str
```

总结:日志能够帮助用户深入学习 OpenStack 和排查问题。但要想高效使用日志还有个前提:

必须先掌握 OpenStack 的运行机制，然后有针对性地查看日志。对于 OpenStack 的运维来说，在大部分情况下，用户都不需要看源代码。因为 OpenStack 的日志记录得很详细，足以帮助用户分析和定位问题。但还是有一些细节日志没有记录，必要时可以通过查看源代码理解得更清楚。即便如此，日志也会为用户提供源代码查看的线索，不需要大海捞针。

7.5 监控指标

在使用 OpenStack 时，必然时时监控其虚拟机性能，下面介绍一个监控 OpenStack 的利器——Cloud Insight。

使用 Cloud Insight 监控 Openstack 前，需要在 OpenStack 为 Cloud Insight Agent 创建单独的角色，确保 Keystone 模块可以让 Agent 访问指标数据。

使用以下指令为 Cloud Insight 创建角色：

```
openstack role create cloudinsight_monitoring
openstack user create cloudinsight --password my_password --project my_project_name
openstack role add cloudinsight_monitoring --project my_project_name --user cloudinsight
```

由以上指令即可创建 Cloudinsight_monitoring 的角色。

用户可以根据实际情况，来修改"my_password"和"my_project_name"。但请注意之后的指令，需将相应的字段修改过来。

此外需要编辑 OpenStack 中的 3 个模块：Nova、Neutron、Keystone 以让角色：cloudinsight_monitoring 获取相应权限。通常，Nova 的权限文件会在 /etc/nova/policy.json 路径下。打开并新增如下权限：

```
Nova
    - "compute_extension:aggregates",
    - "compute_extension:hypervisors",
    - "compute_extension:server_diagnostics",
    - "compute_extension:v3:os-hypervisors",
    - "compute_extension:v3:os-server-diagnostics",
    - "compute_extension:availability_zone:detail",
    - "compute_extension:v3:availability_zone:detail",
    - "compute_extension:used_limits_for_admin",
    - "os_compute_api:os-aggregates:index",
    - "os_compute_api:os-aggregates:show",
    - "os_compute_api:os-hypervisors",
    - "os_compute_api:os-hypervisors:discoverable",
    - "os_compute_api:os-server-diagnostics",
```

```
    - "os_compute_api:os-used-limits"
```
一条完整的权限配置方法，是需要在权限后跟上角色信息。例如：
```
    - "compute_extension:aggregates": "role:cloudinsight_monitoring"
```
根据以上列举的权限和方法，逐个添加权限。若修改过角色名称 cloudinsight_monitoring，请修改角色名称为用户自定义的即可。

Neutron 的权限同理可以使用以上的方法，添加如下权限。而权限配置文件一般在 /etc/neutron/policy.json 中。
```
Neutron
    - "get_network"
```
Keystone 的权限同理可以使用以上的方法，添加如下权限。而权限配置文件一般在 /etc/keystone/policy.json。
```
Keystone
    - "identity:get_project"
    - "identity:list_projects"
```
保存所有的 policy.json 文件后，需要重启 Keystone、Neutron 和 Nova API 来应用新的权限配置。

接下来配置 Cloudinsight Agent 连接到 Keystone，前往 Cloudinsight Agent 的平台服务配置文件中，新增 OpenStack 相关配置。切换路径至 /etc/cloudinsight-agent：
```
cd /etc/cloudinsight-agent
```
开启配置文件 conf.d/openstack.yaml：
```
cp conf.d/openstack.yaml.example conf.d/openstack.yaml
```
修改 keystone_server_url 配置项。通常来说，默认的端口号为 5000。
```
init_config:
  keystone_server_url: "https://my-keystone-server.com:port/"
```
除此之外，还需要前往 <yourHorizonserver>/identity 查询用户的 Project ID 来修改配置项 id。以下是示例：
```
instances:
  - name: instance_1 # A required unique identifier for this instance
    # The authorization scope that will be used to request a token from Identity API v3
    # The auth scope must resolve to 1 of the following structures:
    # {'project':{'name':'my_project', 'domain':'my_domain'} OR {'project':{'id':'my_project_id'}}
    auth_scope:
      project:
        id: b9d363ac9a5b4cceae228e03639357ae
```
前往配置项的 # User credentials 部分，修改角色的账户和密码：

```
# User credentials
# Password authentication is the only auth method supported right now
# User expects username, password, and user domain id
# 'user' should resolve to a structure like {'password': 'my_password', 'name': 'my_name', 'domain'$
user:
  password: my_password
  name: cloudinsight
  domain:
    id: default
```
若在第一步修改过角色的默认账户"cloud insight"和密码"my_password",请在此处对应修改。

配置 RabbitMQ 服务,使用以下指令安装 RabbitMQ 监控插件:

```
rabbitmq-plugins enable rabbitmq_management
```

重启 Rabbit 使插件生效。安装成功后,插件为用户创建 URL 为 http://localhost:15672/api/ 来展示指标数据,而 Cloud Insight 也是通过该地址采集指标数据。

```
service rabbitmq-server restart
```

在 etc/conf.d/rabbitmq.yaml 中添加该地址:

```
instances:
  rabbitmq_api_url: http://localhost:15672/api/
  rabbitmq_user: guest # defaults to 'guest'
  rabbitmq_pass: guest # defaults to 'guest'
```

若 RabbitMQ 已经自定义 API URL 或者用户名和密码,请修改相应配置项。

重启 Cloudinsight Agent,使配置生效:

```
/etc/init.d/cloudinsight-agent restart
```

也可以通过查看 Agent Info 信息,来验证配置是否成功:

```
/etc/init.d/cloudinsight-agent info
```

当出现以下信息,则代表安装成功:

```
Checks
======
  [...]
  openstack
  ------
    - instance #0  [OK]
    - Collected 8 metrics & 0 events
```

Cloud Insight 采集 OpenStack 的性能指标如表 7-1 所示。

表 7-1　性能指标表

指标	单位	具体含义
openstack.nova.current_workload		Hypervisor 当前运行的任务数量
openstack.nova.disk_available_least	gibibytes	Hypervisor 剩余可用磁盘空间，默认单位 GB
openstack.nova.free_disk_gb	gibibytes	Hypervisor 当前可用的磁盘空间，默认单位 GB
openstack.nova.free_ram_mb	mebibytes	Hypervisor 当前可用的 RAM 大小，默认单位 MB
openstack.nova.hypervisor_load.1		Hypervisor 在过去 1min 内的平均负载
openstack.nova.hypervisor_load.15		Hypervisor 在过去 15min 内的平均负载
openstack.nova.hypervisor_load.5		Hypervisor 在过去 5min 内的平均负载
openstack.nova.limits.max_image_meta		分配给租户的最大 image metadata 定义数
openstack.nova.limits.max_personality		分配给租户的最大 personality 数量
openstack.nova.limits.max_personality_size		分配给租户的最大单个 personality 大小
openstack.nova.limits.max_security_group_rules		分配给租户的最大安全组规则数量
openstack.nova.limits.max_security_groups		分配给租户的最大安全组数量
openstack.nova.limits.max_server_meta		分配给租户的最大 server metadata 定义数
openstack.nova.limits.max_total_cores		分配给租户的最大 core 数量
openstack.nova.limits.max_total_floating_ips		分配给租户的最大浮动 IP 数量
openstack.nova.limits.max_total_instances		分配给租户的最大 instance 数量
openstack.nova.limits.max_total_keypairs		分配给租户的最大 keypair 数量
openstack.nova.limits.max_total_ram_size	gibibytes	分配给租户的最大 RAM 大小
openstack.nova.limits.total_cores_used		租户当前使用的 core 数量
openstack.nova.limits.total_floating_ips_used		租户当前使用的浮动 IP 数量
openstack.nova.limits.total_instances_used		租户当前使用 instance 数量
openstack.nova.limits.total_ram_used	gibibytes	租户当前使用 RAM 大小
openstack.nova.limits.total_security_groups_used		租户当前使用的安全组总数
openstack.nova.local_gb	gibibytes	Hypervisor 主机当前临时磁盘大小，默认单位 GB
openstack.nova.local_gb_used	gibibytes	Hypervisor 主机当前磁盘使用量，默认单位 GB
openstack.nova.memory_mb	mebibytes	Hypervisor 主机当前 RAM 大小，默认单位 MB
openstack.nova.memory_mb_used	mebibytes	Hypervisor 主机当前 RAM 使用量，默认单位 MB
openstack.nova.running_vms		Hypervisor 主机当前运行的虚拟机数量
openstack.nova.server.cpu0_time	nanoseconds	虚拟 CPU 的 CPU 时间，默认单位 ns
openstack.nova.server.hdd_errors		Sever 访问 HDD 设备时看到的错误数
openstack.nova.server.hdd_read	bytes	Sever 从 HDD 设备读取的字节数
openstack.nova.server.hdd_read_req		Sever 向 HDD 设备发出的读请求数
openstack.nova.server.hdd_write	bytes	Sever 写入 HDD 设备的字节数

续上表

指标	单位	具体含义
openstack.nova.server.hdd_write_req		Sever 向 HDD 设备发出的写请求数
openstack.nova.server.memory	mebibytes	为当前 sever 配置的内存量
openstack.nova.server.memory_actual	mebibytes	为当前 sever 配置的内存量
openstack.nova.server.memory_rss	mebibytes	Sever 进程当前使用的内存量，即与磁盘页面没有关联（如堆栈和对内存）
openstack.nova.server.vda_errors		Sever 访问 VDA 设备时的错误数
openstack.nova.server.vda_read	bytes	Sever 从 VDA 设备读取的字节数
openstack.nova.server.vda_read_req		Sever 向 VDA 设备发出的读请求数
openstack.nova.server.vda_write	bytes	Sever 写入 VDA 设备的字节数
openstack.nova.server.vda_write_req		Sever 向 VDA 设备发出的写请求数
openstack.nova.vcpus		Hypervisor 主机的 vCPU 配额
openstack.nova.vcpus_used		Hypervisor 主机当前使用的 vCPU 数量

7.6 备份与恢复

一般在设计 OpenStack 时就应该考虑系统的备份策略。系统备份注意事项如下：
① 要保留多少备份？
② 是否需要异地备份？
③ 备份间隔多久？
④ 恢复策略？

7.6.1 备份的分类

1. 数据库备份

制定 crond 每天备份一次。备份命令根据系统而定，可使用 mysqldump 或者 xtrabackup 命令。

2. 文件系统备份

（1）计算服务。

备份文件：/etc/nova、/var/lib/nova、/var/log/nova（已做日志服务器不用备份）。

其中，/var/lib/nova/instances 一般无须备份，此目录为虚拟机存放目录，一般为共享目录，或将虚拟机存放于后端存储。

备份运行中的 KVM 实例，还原后有可能导致虚拟机无法引导。

（2）镜像目录。

备份文件：/etc/glance、/var/log/glance（已做日志服务器不用备份）、/var/lib/glance。

其中，/var/log/glance/images 无须备份，此目录和 nova 实例一样一般为共享目录或者存放在后端存储，如果本目录是文件系统，需要另行备份。

可直接利用 rsync 或者 scp 定期复制到备份服务器上。

(3) 身份服务。

备份文件：/etc/keystone、/var/log/keystone、/var/lib/keystone（此目录不包含再用数据，选择备份）。

(4) 块存储。

备份文件：/etc/cinder、/var/log/cinder、/var/lib/cinder。

(5) 对象存储。

备份文件：/etc/swift、swift 配置文件、环文件、环生成文件。

一般将环文件和环生成文件复制到所有的存储结点中，这样会存放多个文件副本，单点故障无影响。

(6) 恢复备份。

恢复 nova 步骤如下：

①停止 nova 的所有服务。

②恢复 nova 的数据库。

③备份当前文件：mv /etc/nova{, .orig}。

④恢复文件：cp -a backup/nova /etc/。

⑤启动 nova 进程。

其他组件恢复步骤一致。

7.6.2 MySQL 数据库备份与恢复

首先，需要一个 MySQL 数据库备份脚本。需要注意的是：OpenStack 的 MySQL 数据库中的 root 账户的密码是随机设置的，其值在 /etc/contrail/mysql.token 中。命令如下：

```bash
#!/bin/bash
backup_dir="/opt/backup/mysql"
if [ ! -d "$backup_dir" ]; then
  mkdir -p "$backup_dir"
fi
# Dump the entire mysql
/bin/nice -n 19 /usr/bin/mysqldump -uroot -p`cat /etc/contrail/mysql.token` --opt --flush-logs --single-transaction --ignore-table=mysql.event --ignore-table=mysql.gtid_slave_pos --ignore-table=mysql.innodb_index_stats --ignore-table=mysql.innodb_table_stats --all-databases > ${backup_dir}/mysql-`hostname`-`eval date +%Y%m%d`.sql
    /bin/nice -n 19 tar zPcf ${backup_dir}/mysql-`hostname`-`eval date +%Y%m%d`.sql.tar.gz ${backup_dir}/mysql-`hostname`-`eval date +%Y%m%d`.sql
    rm -rf $backup_dir/*.sql
```

然后，要将这个备份脚本设置为自动运行，如设置为每天凌晨三点执行，命令如下：

```
echo "0 03 * * * root /usr/bin/sh /opt/backup/shell/backupmysql.sh" >> /etc/crontab
```

最后，一旦由于异常断点等原因导致数据库文件丢失，MySQL 服务启动失败，则安装下列步骤进行恢复：

① 正常情况下，一个刚 yum 装完的数据库目录下有这些文件。

```
[root@test mysql]# ll
total 28
-rw-rw---- 4 mysql mysql 16384 Aug 29 13:35 aria_log.00000001
-rw-rw---- 4 mysql mysql    52 Aug 29 13:35 aria_log_control
drwx--x--x 2 mysql mysql  4096 Dec 21 14:33 mysql
drwx------ 2 mysql mysql  4096 Dec 21 14:33 performance_schema
drwxr-xr-x 2 mysql mysql     6 Aug 29 13:35 test
```

② 对比当前的 /var/lib/mysql/ 文件夹下的文件，把多余的文件都删除。
③ 启动数据库 service mysql start。
④ 导入数据库文件，恢复数据。

```
# cat /etc/contrail/mysql.token
f0d330d601ce33f1a69f
# mysql -u root -p < /mysql-flexhcs_controller_1-20161221.sql
Enter password: f0d330d601ce33f1a69f
```

7.7 OpenStack 常用故障处理方法

对于初学者，在进行 OpenStack 运维时经常会遇到各种各样的故障，以下是运维过程中一些常见故障及处理方法的汇总，可供读者参考。

1. 故障处理流程

首先确定故障的资源 ID，并判断故障发生的组件。

查看对应组件的日志，根据故障资源 ID 进行搜索，找到相应的 ERROR 日志。

如果 ERROR 日志又将问题指向了其他组件，则根据 ERROR 日志中的资源 ID、时间、req-id 等信息，其他组件继续查找问题，直到发现问题根源。

如果没有在日志中发现任何 ERROR，有可能是网络不通，导致请求无法到达 API，此时需要检查和 API 的连通性（如果使用 VIP，需要分别检查和 VIP 的联通性和实 IP 的连通性）。

如果 API 中能找到相应请求，但是 conductor/scheduler/compute 没有找到相应日志，则有可能是 MQ 发生故障。

如果组件长时间没有任何日志刷新，有可能是组件进程挂掉或者处于僵死状态，可以尝试重启服务，或先打开 Debug 再重启服务。

2. 创建 vm 报错

```
cat nova-compute.log | grep 620cd801-8849-481a-80e0-2980b6c8dba6
 2019-02-23 15:13:36.136 3558 INFO nova.compute.resource_tracker
[req-f76d5408-00f8-4a67-854e-ad3da2098811 - - - -] Instance 620cd801-
```

8849-481a-80e0-2980b6c8dba6 has allocations against this compute host but is not found in the database.

分析：似乎是 node 的信息数据库不同步，nova show 出错的 vm，包 cell 错误。

方法：每次增加一个计算结点在控制结点需要执行。

su -s /bin/sh -c "nova-manage cell_v2 discover_hosts --verbose" nova

3. 查看中文命名的镜像时报错

glance image-list

'ascii' codec can't encode character u'\u5982' in position 1242: ordinal not in range(128)

分析：镜像名字有用中文命名的情况。

方法：将 export LC_ALL=zh_CN.UTF-8 添加到 /etc/profile，同时注意 source 的文件中是否有 export LC_ALL=C。

4. 无法显示控制台

操作步骤：控制台 - 虚拟机，单击虚拟机名称，单击"控制台"。

预期结果：正常显示 console 页面。

实际结果：页面提示 "Failed to connect to server"，单击可以在新窗口打开 console 页面。

方法：nova 控制结点的配置问题，其配置文件中 memcache 以及 rabbitmq 的 hosts 配置不对。

5. import error

故障：glance register.log 报错。

2018-05-18 03:18:55.890 3185 ERROR glance.common.config [-] Unable to load glance-registry-keystone from configuration file /usr/share/glance/glance-registry-dist-paste.ini.

Got: ImportError('No module named simplegeneric',)

方法：/usr/lib/python2.7/site-packages/simplegeneric.py 没有读权限，被谁给改了

6. Keystone 报错

systemctl status httpd.service

httpd.service - The Apache HTTP Server
   Loaded: loaded (/usr/lib/systemd/system/httpd.service; enabled; vendor preset: disabled)
   Active: failed (Result: exit-code) since Sat 2016-05-28 20:22:34 EDT; 11s ago
     Docs: man:httpd(8)
           man:apachectl(8)
  Process: 4501 ExecStop=/bin/kill -WINCH ${MAINPID} (code=exited, status=1/FAILURE)
  Process: 4499 ExecStart=/usr/sbin/httpd $OPTIONS -DFOREGROUND (code=exited, status=1/FAILURE)

Main PID: 4499 (code=exited, status=1/FAILURE)
May 28 20:22:34 controller0 httpd[4499]:(13)Permission denied: AH00072:make_sock:could not bind to address [::]:5000
May 28 20:22:34 controller0 httpd[4499]:(13)Permission denied: AH00072:make_sock:could not bind to address 0.0.0.0:5000

故障：Permission denied：AH00072：make_sock：could not bind to address [::]：5000

方法：可能是防火墙的配置问题或者 SELinux，若防火墙没有问题，检查 SELinux 状态。

```
getenforce
enforcing  # 如果不为 disabled 则表示为 selinux 正常运行
SELINUX=enforcing 改为 selinux=distabled
```

重启 reboot

7. Dashboard 访问卡顿

定位到 Dashboard 卡顿的原因是因为 Nova 卡顿，Nova 卡顿是因为 Nova 无法与 Memcached 建立连接，进一步定位到 Memcached 默认的最大连接数是 1 024，目前已达到最大连接数。

解决办法为编辑 /etc/sysconfig/memcached。

参数修改为：

```
PORT="11211"
USER="memcached"
MAXCONN="65536"
CACHESIZE="1024"
OPTIONS=""
```

重启 memcached

8. Access denied for user 'nova'@'%' to database 'nova_api'

故障：初始化 nova_api 数据库时出现。

```
su -s /bin/sh -c "nova-manage api_db sync" nova
```

报错：

```
Access denied for user 'nova'@'%' to database 'nova_api'
```

方法：root 用户进入数据库，执行。

```
> GRANT ALL PRIVILEGES ON nova_api.* TO 'nova'@'%' IDENTIFIED BY '2267593eb27be7c414fc';
```

9. novnc 打不开问题定位

分析：可能 compute 的防火墙被修改。

方法：在 /etc/sysconfig/iptables 添加。

```
-A INPUT -p tcp --dport 5900:6100 -j ACCEPT
```

10. 并发创建虚机，有部分失败

分析：在 log 中看到错误为 rabbitmq 端口连接不上，初步定位到是由于 rabbitmq 压力过大，导致 rabbitmq 不能正常工作。可采取一些措施来改善此情况。

方法：把 rabbitmq 的集群的 rabbitmq 结点 2 和结点 3 由 disc 模式改为 ram 模式。把 rabbitmq

压力分散到 3 个 rabbitmq 结点。通过查看 rabbitmq log 发现在没有修改之前，rabbitmq 压力主要在 node01 上，其他两个结点几乎不会处理消息请求。把 rabbitmq 监控的 rate mode 功能关闭。这个 rate mode 功能监控消息队列中消息传输速率。关闭对服务没有影响。Nova-compute 之前配置的 [glance] api_servers=vip：9292，vip 是管理网地址，当创建虚拟机等并发大时，会出现镜像下载占用管理网通道，导致 rabbitmq 等网络敏感服务消息阻塞，甚至消息超时，因此，将其 api_servers 配置为控制结点的存储地址，且把控制结点镜像服务由监听管理网改为监听存储网。

11. 一台控制结点内存使用率过高报警

一台控制结点内存使用率过高报警，发现是 rabbitmq 进程异常导致，消息队列中积压的消息过多导致内存增大无法释放，重启 rabbitmq 进程解决问题，最终解决问题需要修改 rabbitmq 配置文件，使得积压的消息存储在磁盘中而不是内存中。

12. 创建 vm 时报 host doesn't support passthrough 错误

分析：这个错误有可能由于母机 VT-d（或 IOMMU）未开启导致。确保 "intel_iommu=on" 启动参数已经按上文叙述开启。发现已经修改了 /etc/default/grub 文件。

方法：配置计算结点的 /etc/default/grub 文件，在 GRUB_CMDLINE_LINUX 中添加 intel_iommu=on 来激活 VT-d 功能，重启物理机。

```
$ cat /etc/default/grub
GRUB_TIMEOUT=5
GRUB_DISTRIBUTOR="$(sed 's, release .*$,, g' /etc/system-release)"
GRUB_DEFAULT=saved
GRUB_DISABLE_SUBMENU=true
GRUB_TERMINAL_OUTPUT="console"
GRUB_CMDLINE_LINUX="crashkernel=auto rd.lvm.lv=bclinux/root rd.lvm.lv=bclinux/swap intel_iommu=on rhgb quiet"
GRUB_DISABLE_RECOVERY="true"
```

【注意】不重启不生效，重启后可以正常生成 sriov 虚拟机。

对于 intel 的 cpu 和 amd 的 cpu，在 grub 配置上是不同的，具体的配置请参考文章：http://pve.proxmox.com/wiki/Pci_passthrough。

在编辑完 grub 文件后，需要更新。

```
grub2-mkconfig      # fedora arch centos
update-grub         # ubuntu debian
```

重启计算机，使其生效。

```
# cat /proc/cmdline
```

单元 8 综合案例

学习目标

◎ 掌握使用 Python 对 OpenStack 进行二次开发，对镜像、云主机类型、云主机运行进行管理和功能添加等；

◎ 熟悉 OpenStack API 的使用。

8.1 OpenStack 数据库详解

8.1.1 项目相关数据库

本单元主要讲解利用 Python 对 OpenStack 进行二次开发的代码部分，OpenStack 提供 API 接口，另外，可以直接访问 OpenStack 相关数据库，这为二次开发提供了基础，这里主要使用其中三个数据库，分别是 glance、nova、nova_api。

1. 数据库：glance

数据表：images。

作用：存储所有的镜像信息，如表 8-1 所示。

表 8-1 镜像信息

序号	字段名	类型	能否为空	主键	字段说明
1	id	varchar(36)	NO	PRI	主键
2	name	varchar(255)	YES		镜像名称
3	size	bigint(20)	YES		镜像大小
4	status	varchar(30)	NO		当前镜像状态
5	created_at	datetime	NO		创建时间
6	updated_at	datetime	YES		最后更新时间

续上表

序号	字段名	类型	能否为空	主键	字段说明
7	deleted_at	datetime	YES		删除时间
8	deleted	tinyint	NO		是否已删除（1 为已删除，0 为未删除）
9	disk_format	varchar(20)	YES		镜像格式

2. 数据库：nova_api

数据表：flavors。

作用：云主机类型表，存储 OpenStack 云主机类型的信息，如表 8-2 所示。

表 8-2　云主机类型信息

序号	字段名	类型	能否为空	主键	字段说明
1	created_at	datetime	YES		创建时间
2	updated_at	datetime	YES		更新时间
3	name	varchar(255)	NO		云主机类型名称
4	id	int	NO	PRI	主键
5	memory_mb	int	NO		内存大小
6	vcpus	int	NO		CPU 个数
7	flavorid	varchar(255)	NO		云主机类型 id
8	root_gb	int	YES		磁盘大小

3. 数据库：nova

数据表：instances。

作用：云主机信息表，存储 OpenStack 所有云主机的信息，如表 8-3 所示。

表 8-3　云主机信息信息

序号	字段名	类型	能否为空	主键	字段说明
1	created_at	datetime	YES		创建时间
2	updated_at	datetime	YES		更新时间
3	deleted_at	datetime	YES		删除时间
4	id	int	NO	PRI	主键
5	image_ref	varchar(255)	YES		引用镜像 id
6	vm_state	varchar(255)	YES		当前状态
7	memory_mb	int	YES		内存大小
8	vcpus	int	YES		CPU 个数
9	hostname	varchar(255)	YES		云主机名称
10	host	varchar(255)	YES		所属结点
11	uuid	varchar(36)	NO		云主机 id
12	root_gb	int	YES		磁盘大小

8.1.2 OpenStack 中主要的数据库表

下面是 OpenSatck 数据库中其他主要表，这里只简单介绍其主要作用，具体的表结构可以自行查看。

1. 数据库 keystone

（1）endpoint：各个模块的访问地址。
（2）migrate_version：迁移版本。
（3）service：卷服务。

2. 数据库 glance

（1）images：镜像表。
（2）migrate_version：迁移版本。

3. 数据库 nova

（1）compute_nodes：存放各个计算结点的信息，包括多少个虚拟机。service_id 对应 service 表中的 compute 的 id。
（2）fixed_ip：fixed 的分配情况。
（3）floating_ip：外网 ip 的分配情况。
（4）instance_faults：记录与实例有关的错误。
（5）instance_info_caches：与实例的网络设置缓存有关。
（6）instance_types：与 flavor 设置有关。
（7）instances：实例主表，存储所有云主机信息。
（8）key_pairs：密钥。
（9）security_group_instance_association：实例与安全组的关联表。
（10）security_group_rules：安全组角色。
（11）security_groups：安全组。
（12）services：控制结点和计算结点所有的服务。

4. 数据库 nova_api

flavors：云主机类型信息。

8.2 OpenStack API 理解

8.2.1 使用 OpenStack 服务的方式

OpenStack 项目作为一个 IaaS 平台，提供以下三种使用方式：

（1）通过 Web 界面，也就是通过 Dashboard（面板）来使用平台上的功能。
（2）通过命令行，也就是通过 keystone、nova、neutron 等命令，或者通过最新的 OpenStack 命令来使用各个服务的功能（社区目前的发展目标是使用一个单一的 OpenStack 命令替代过去的每个项目一个命令的方式，以后会只存在一个 OpenStack 命令）。

(3) 通过 API，也就是通过各个 OpenStack 项目提供的 API 来使用各个服务的功能。

在上面提到的三种方式中，通过 API 这种方式是基础，是其他两种方式可行的基础。

通过 Web 界面是用 OpenStack 服务这种方式由 OpenStack 的 Horizon 项目提供。Horizon 项目是一个 Django 应用，实现了一个面板功能，包含前后端的代码（除了 Python，还包括 CSS 和 JS）。Horizon 项目主要是提供一种交互界面，它会通过 API 来和各个 OpenStack 服务进行交互，然后在 Web 界面上展示各个服务的状态；它也会接收用户的操作，然后调用各个服务的 API 来完成用户对各个服务的使用。

通过命令行是用 OpenStack 服务的方式，由一系列项目提供，这些项目一般都命名为 python-projectclient，如 python-keystoneclient、python-novaclietn 等。这些命令行项目分别对应各个主要的服务，为用户提供命令行操作界面和 Python 的 SDK。例如：python-keystoneclient 对应到 Keystone，为用户提供 Keystone 这个命令，同时也提供 keyston 项目的 SDK（其实是在 SDK 的基础上实现命令行）。这些 Client 项目提供的 SDK 其实也是封装了对各自服务的 API 的调用。由于每个主要项目都有一个自己的命令行工具，社区觉得不好，于是又增加了一个新的项目 python-OpenStackclient，用来提供一个统一的命令行工具 OpenStack（命令的名字就是 OpenStack），这个工具实现了命令行，然后使用各个服务的 Client 项目提供的 SDK 来完成对应的操作。

通过 API 是用 OpenStack 的方式由各个服务自己实现，如负责计算的 Nova 项目实现了计算相关的 API，负责认证的 Keystone 项目实现了认证和授权相关的 API。这些 API 都是有统一的形式的，都是采用了 HTTP 协议实现的符合 REST 规范的 API。

8.2.2　OpenStack 中的常见 API 类型

在 OpenStack 中不管程序内部之间的调用，还是对于 OpenStack 中的各种服务和功能的内部调用，或者是外部调用均通过 API 的形式来进行。这里分析一下 OpenStack 中的几种常见 API 类型。

(1) 程序内部的 API 主要是给本机程序内部使用，如 nova_master/nova/compute/api.py 文件中的 API Class 主要是为了给 manager 调用，其中调用哪个 API Class 也是利用 OpenStack 中非常重要的动态载入方法来确定的，非常灵活。

(2) API 是 RPC API，就是通过高级消息队列的方式实现不同主机的方法的远程调用，如 nova_master/nova/compute/rpcapi.py，其中调用的方法都是 manager 中的方法。通过 RPC 的方式是实现分布式程序的基本方法，采用消息队列的 RPC 方式是目前流行的多种云计算框架实现的普遍方式。

(3) API 就是通过 Web 资源的方式暴露给外界的 API，将提供的服务暴露成 Web 资源，可以方便外界访问，OpenStack 是通过其一个对应一类 API 的 WSGIService 服务来实现对外的服务。

(4) API 就是 Client API，是对 Web API 的封装，提供这种形式的 API 主要是方便用户对复杂的 Web 资源形式的 API 的调用，简化操作，便于用户通过程序调用。

8.2.3　OpenStack API 使用规范

OpenStack 的各个服务组件都提供相应的 API 接口，如 nova-api、gance-api 等。使用这些 API

和扩展可以让用户在 OpenStack 云中进行一系列的操作，如启动服务主机、创建镜像、给主机和镜像设置元数据、创建容器和对象等。

但在调用任何组件的 API 前，必须通过 Identity API 进行鉴权。

1. Identity Service 身份认证

图 8-1 所示是一个权限验证的过程。

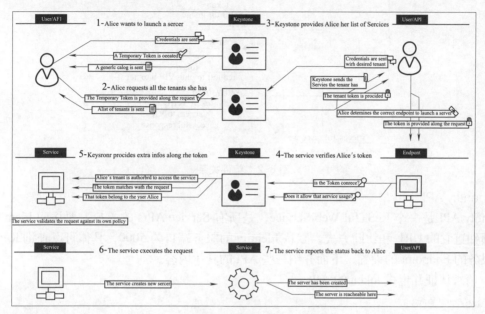

图 8-1　权限验证

OpenStack 身份处理流程（OpenStack Identity process flow）：

（1）用户想要创建一个服务器，首先要提供自己的认证信息（可能是用户名和密码，或者是用户名和 API key），然后 Keystone 提供一个临时的 token（令牌）和一个服务列表（ServiceCatalog）。

（2）用户使用临时的 token 请求他所有的 Tenant 信息（可以理解为一些资源），Keystone 返回 Tenants 列表。

（3）Keystone 提供用户目标 Tenant 里面的服务列表。用户用自己的鉴权信息获得目标 Tenant 的 token 和服务列表，然后使用 Tenant token 和指定服务的 Endpoint（可以理解为具体的 API）信息去启动服务器。

（4）该服务与 Keystone 进行验证该 token 是否正确。

（5）Keystone 返回这个 token 的相关信息：用户获得鉴权被允许访问这个服务；token 和这个请求相符；token 属于这个用户。

（6）服务执行该请求（创建服务器）。

（7）执行完成后返回结果。

OpenStack 中调用任何 API，在执行请求前都要经过 Keystone 认证，图 8-2 说明了 Keystone 和其他各个服务组件的关系。

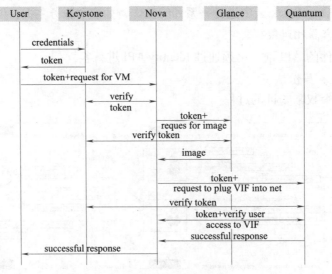

图 8-2　组件关系

2. Identity API

Identity API 是一个 ReSTful Web Service，是所有 Service APIs 的入口。要访问 Identity API，用户必须知道它的 URL 和访问方式。默认 Identity 的服务端口为 5000，具体可以查询 Keystone 中对各个服务的 Endpoint 的设置。下面介绍一些 API 的具体内容。

(1) 请求认证并生成 token 的 API：

method：'POST'

URL：'/v2.0/tokens'

(2) 每个 ReST 请求需要有 X-Auth-Token 作为 header。

(3) 认证时，需要提供 user ID 和密码或者是一个 token。

如果 token 过期，会返回 401；如果是请求中的 token 过期，会返回 404；Identity 将过期 token 视为无效 token。

(4) 部署时可以指定 token 的过期时间。

(5) 正常返回码：200，203。

错误返回码：identityFault（400，500，…），userDisabled（403），badRequest（400），unauthorized（401），forbidden（403），badMethod（405），overLimit（413），serviceUnavailable（503），itemNotFound（404）。

(6) 所有的 API 提供 JSON 和 XML 两种格式。

读者可以访问 http：//developer.openstack.org/api-ref.html 官方文档获取更具体的内容。

8.3　获取镜像列表

在开始编写功能代码之前需要先编写 OpenStack API 的调用类（api_class.py），用来初始化 tocken，连接 OpenStack。所有的地址拼接等信息都在 api_class.py 中。

命令代码如下：

```python
class stack_api:
    OS_AUTH_URL='http://192.168.0.190'  # IP 地址
    headers={'Accept': '*/*', 'Connection': 'close'}
    tokes=None
    headers['Content-Type']='application/json'

    # 初始化 token（令牌）
    def __init__(self):
        stack_api.tokes=self.get_token()

    # 获取 token（认证令牌）
    def get_token(self):
        body={'auth':
                {'identity':
                    {'methods': ['password'],
                     'password': {
                         'user': {
                             'domain': {'name': 'default'},
                             'name': 'admin',       # 用户名
                             'password': '******'   # 密码
                         }
                     }
                    },
                 'scope': {
                     'project': {
                         'domain': {'name': 'default'},  # 默认域
                         'name': 'admin'}
                     }
                }
             }
        get_token_url=stack_api.OS_AUTH_URL + ':35357/v3/auth/tokens'
        result=requests.post(get_token_url, headers=stack_api.headers,
            data=json.dumps(body)).headers['X-Subject-Token']
        return result
```

获取镜像列表
(1) api_calss.py 中的代码如下：

```python
def get_images_list(self):
    images_list_url=stack_api.OS_AUTH_URL + ':9292/v2/images'
    stack_api.headers['X-Auth-Token']=stack_api.tokes
    images_list=requests.get(images_list_url, headers=stack_api.headers).json()
    return images_list
```

再创建一个模块（method.py）来对时间进行转换：

```python
def utc2local(utc_st):
    UTC_FORMAT="%Y-%m-%dT%H:%M:%SZ"
    utc_=datetime.datetime.strptime(utc_st, UTC_FORMAT)
    now_stamp=time.time()
    local_time=datetime.datetime.fromtimestamp(now_stamp)
    utc_time=datetime.datetime.utcfromtimestamp(now_stamp)
    offset=local_time - utc_time
    local_st=str(utc_ + offset)
    return local_st
```

(2) 具体获取的逻辑在 app.py 中，代码如下：

```python
@app.route('/get_image_data', methods=['GET', 'POST'])
def return_image_data():
    images_list=obj.get_images_list()
    for i in images_list["images"]:
        i["created_at"]=method.utc2local(str(i["created_at"]))
        i["updated_at"]=method.utc2local(str(i["updated_at"]))
    return json.dumps(images_list)
```

(3) 后台代码写完后就是页面代码，部分 JS 代码如下（image_manage.html）：

```javascript
$(function () {
// 清空所有的子结点
$("#J_TbData").empty();
// 加载镜像数据
$.ajax({
    url: "/get_image_data",// 后台请求的数据
    dataType: "json",// 数据格式
    type: "get",// 请求方式
    async: false,// 是否异步请求
    success: function (data) {    // 如果请求成功，返回数据。
        var image_list=data["images"];      // 镜像相关数据
        for (var j=0; j < image_list.length; j++) {
            var i=image_list[j];
```

```
            // 动态创建一个 tr 行标签，并且转换成 jQuery 对象
            var $trTemp=$("<tr></tr>");
            // 往行里面追加 td 单元格
            $trTemp.append("<td>" + "<input type='checkbox' name='foo'
class='id' value='" + i["id"] + "'>" + "</td>");
            $trTemp.append("<td>" + i["name"] + "</td>");
            $trTemp.append("<td>" + "镜像" + "</td>");
            $trTemp.append("<td>" + i["status"] + "</td>");
            $trTemp.append("<td>" + i["disk_format"] + "</td>");
            $trTemp.append("<td>" + (i["size"] / 1048576).toFixed(2) +
"M" + "</td>");
            $trTemp.append("<td>" + i["created_at"] + "</td>");
            $trTemp.append("<td>" + i["updated_at"] + "</td>");
            $trTemp.append("<td>" + "<input type='button' class=
'created' value='创建云主机'>" + "<input type='button'
class='update' value='修 改'>" + "<input type='button'
class='del' value='删除'>" + "</td>");
            $trTemp.appendTo("#J_TbData");
        }
    },
});
```

8.4 镜像上传与编辑

8.4.1 创建镜像（镜像上传）

创建镜像部分采用上传本地镜像的方式，需要选择本地文件，镜像格式采用下拉选择的方式。
(1) api_class.py 部分代码如下：

```
def image_upload(self, name, disk_format, image_file,
visibility=None, min_disk=None, min_ram=None, protected=None,
container_format='bare'):
    request_url=stack_api.OS_AUTH_URL + ':9292/v2/images'
    bodys={"disk_format": disk_format, "name": name}
    if visibility is not None:
        bodys["visibility"]=visibility
    if min_disk is not None:
        bodys["min_disk"]=min_disk
    if min_ram is not None:
```

```
        bodys["min_ram"]=min_ram
    if protected is not None:
        bodys["protected"]=protected
    if container_format is not None:
        bodys["container_format"]=container_format
    stack_api.headers['X-Auth-Token']=stack_api.tokes
    result=requests.post(request_url, headers=stack_api.headers,
    data=json.dumps(bodys)).json()
    file_url=result['file']    # 镜像上传地址
    stack_api.headers['Content-Type']='application/octet-stream'
    image_upload_url=stack_api.OS_AUTH_URL + ':9292' + file_url   # 拼接
完整 http 请求地址
    requests.put(image_upload_url, headers=stack_api.headers,
    data=image_file)
```

(2) app.py 中的部分代码如下：

```
if request.method=="POST":
file=request.files.get("file")
name=request.form.get("name")
format=request.form.get("format")
obj.image_upload(name, format, file)
return redirect(url_for("images"))
elif request.method=="GET":
return render_template("image_manage.html", net_list=net_list,
flavor_list=flavor_list)
```

这里的镜像上传采用的是 form 表单提交的方式，这里不再具体解释。

8.4.2 删除镜像

(1) 删除镜像是根据镜像 id 来删除，api_calss.py 部分代码如下：

```
def delete_image(self, image_id):
request_url=stack_api.OS_AUTH_URL + ':9292/v2/images/' + image_id
stack_api.headers['X-Auth-Token']=stack_api.tokes
requests.delete(request_url, headers=stack_api.headers)
```

(2) 参数拼接完成后就调用方法来删除，app.py 中的部分代码如下：

```
elif request.method=="DELETE":
image_id=json.loads(request.form.get("image_id"))
for i in range(len(image_id)):
    obj.delete_image(image_id[i])
return "success"
```

(3) 前台页面的代码如下（image_manage.html）：

```
function del(data) {
```

```
$.ajax({
    url: '/image_manage',
    type: 'delete',
    dataType: "text",
    data: {"image_id": JSON.stringify(data)},
    success: function (text) {
        if (text=="success") {
            alert("删除成功")
            location.href="image_manage";
        }
    }
});
}
```

8.4.3 镜像修改

镜像修改只是修改镜像名，这里只是提供一个思路，可以根据需要进行其他操作。

(1) api_class.py 中的部分代码如下：

```
def updata_image(self, image_id, name):
request_url=stack_api.OS_AUTH_URL + ':9292/v2/images/' + image_id
stack_api.headers['Content-Type']='application/openstack-images-v2.1-json-patch'
stack_api.headers['X-Auth-Token']=stack_api.tokes
data=[{"op": "replace", "path": "/name", "value": name}]
requests.patch(request_url, headers=stack_api.headers, data=json.dumps(data))
```

(2) app.py 部分代码如下

```
elif request.method=="GET":
image_id=request.args.get("image_id")
name=request.args.get("name")
obj.updata_image(image_id, name)
return render_template("image_manage.html")
```

镜像修改采用表单提交的方式，不做过多介绍。

8.5 获取云主机列表

(1) api_class.py 中的部分代码如下：

```
def flavor_list(self):
flavor_url=stack_api.OS_AUTH_URL + ':8774/v2/flavors/detail'
```

```
dc={'Accept': '*/*', 'Connection': 'close', 'X-Auth-Token': stack_
api.tokes}
flavor_type=requests.get(flavor_url, headers=dc).json()
return flavor_type
```

（2）app.py 中的部分代码如下：

```
if request.method=="GET":
flavor_list=obj.flavor_list()
return json.dumps(flavor_list)
```

（3）Flavor_manage.html 代码如下：

```
$(function () {
// 清空所有的子结点
$("#J_TbData").empty();
// 加载 flavor 数据
$.ajax({
    url: "/get_flavor_data",// 后台请求的数据
    dataType: "json",// 数据格式
    type: "get",// 请求方式
    async: false,// 是否异步请求
    success: function (data) {     // 如果请求成功，返回数据
        var image_list=data["flavors"];        // 镜像相关数据
        for (var j=0; j<image_list.length; j++) {
            var i=image_list[j];
            // 动态创建一个 tr 行标签，并且转换成 jQuery 对象
            var $trTemp=$("<tr></tr>");
            // 往行里面追加 td 单元格
            $trTemp.append("<td>" + "<input type='checkbox' name=
'foo' class='id' value='" + i["id"] + "'>" + "</td>");
            $trTemp.append("<td>" + i["name"] + "</td>");
            $trTemp.append("<td>" + i["vcpus"] + "</td>");
            $trTemp.append("<td>" + i["ram"] + "M" + "</td>");
            $trTemp.append("<td>" + i["disk"] + "G" + "</td>");
            $trTemp.append("<td>" + i["id"] + "</td>");
            $trTemp.append("<td>" + "<input type='button' class=
'update' value=' 修 改 '>" + "<input type='button'
class='del' value=' 删除 '>" + "</td>");
            $trTemp.appendTo("#J_TbData");
        }
    },
})
```

8.6 云主机相关操作

8.6.1 创建云主机

创建云主机需要云主机名称、flavor、网络。

(1) Api_class.py 部分代码如下：

```python
def cloud_host_create(self, name, imageRef, flavorRef, networks):
    request_url=stack_api.OS_AUTH_URL + ':8774/v2/servers'
    bodys={'server': {}}
    bodys['server']['name']=name
    bodys['server']['imageRef']=imageRef
    bodys['server']['flavorRef']=flavorRef
    bodys['server']['networks']=[{"uuid": networks}]
    dc={'Accept': '*/*', 'Connection': 'close', 'X-Auth-Token': stack_api.tokes}
    requests.post(request_url, data=json.dumps(bodys), headers=dc)
```

(2) app.py 部分代码如下：

```python
if request.method=="POST":
    name=request.form.get("name")
    image_id=request.form.get("image_id")
    flavor=request.form.get("flavor")
    network=request.form.get("network")
    obj.cloud_host_create(name, image_id, flavor, network)
    return redirect(url_for('servers'))
```

image_manage.html 中采用表单提交的方式，这里不做过多介绍。

8.6.2 创建云主机类型

创建云主机类型需要类型名称、VCPU 数量、内存、硬盘相关信息。

(1) api_class.py 部分代码如下：

```python
def flavor_type_create(self, name, ram, disk, vcpus, is_public=None, description=None, id=None):
    flavor_url=stack_api.OS_AUTH_URL + ':8774/v2/flavors'
    bodys={'flavor': {}}
    bodys['flavor']['name']=name
    bodys['flavor']['ram']=ram
    bodys['flavor']['disk']=disk
    bodys['flavor']['vcpus']=vcpus
    if is_public is not None:
```

```
        bodys['flavor']['os-flavor-access:is_public']=is_public
if description is not None:
        bodys['flavor']['description']=description
if id is not None:
        bodys['flavor']['id']=id
stack_api.headers['X-Auth-Token']=stack_api.tokes
requests.post(flavor_url, headers=stack_api.headers, data=json.dumps(bodys))
```

(2) app.py 部分代码如下:
```
def flavors():
# 创建flavor
if request.method=="POST":
        name=request.form.get("name")
        vcpu=int(request.form.get("vcpu"))
        ram=int(request.form.get("ram")) * 1024
        disk=int(request.form.get("disk"))
        obj.flavor_type_create(name, ram, disk, vcpu)
        return render_template("flavor_manage.html")
elif request.method=="GET":
        return render_template("flavor_manage.html")
```
在 flavor_manage.html 中采用 from 表单提交的方式,这里不做过多介绍。

8.6.3 修改云主机类型

修改云主机类型采用的是先删除再创建的方式。

app.py 部分代码如下:
```
elif request.method=="POST":
name=request.form.get("name_")
vcpu=int(request.form.get("vcpu_"))
ram=int(request.form.get("ram_")) * 1024
disk=int(request.form.get("disk_"))
flavor_id=request.form.get("flavor_id")
obj.delete_flavor(flavor_id)
time.sleep(5)
obj.flavor_type_create(name, ram, disk, vcpu)
return render_template("flavor_manage.html")
```
在 flavor_manage.html 中采用 from 表单提交的方式,这里不做过多介绍。

8.6.4 删除云主机类型

同样需要根据 id 来删除。

(1) api_class.py 部分代码如下：

```python
def delete_flavor(self, flavor_id):
    flavor_url=stack_api.OS_AUTH_URL + ':8774/v2/flavors/' + flavor_id
    stack_api.headers['X-Auth-Token']=stack_api.tokes
    requests.delete(flavor_url, headers=stack_api.headers)
```

(2) app.py 部分代码如下：

```python
elif request.method=="DELETE":
    flavor_id=json.loads(request.form.get("flavor_id"))
    for i in range(len(flavor_id)):
        obj.delete_flavor(flavor_id[i])
    return "success"
```

(3) flavor_manage.html 中的删除方法如下：

```javascript
function del(data) {
$.ajax({
    url: '/flavor_manage',
    type: 'delete',
    dataType: "text",
    data: {"flavor_id": JSON.stringify(data)},
    success: function (text) {
        if (text=="success") {
            alert("删除成功");
            location.href="flavor_manage"
        }
    }
});
}
```

删除选中与单个删除的操作是一样的。

8.6.5 云主机运行管理

1. 获取云主机运行信息

(1) api_class.py 部分代码如下：

```python
def cloud_host_list(self):
    request_url=stack_api.OS_AUTH_URL + ':8774/v2/servers/detail'
    dc={'Accept': '*/*', 'Connection': 'close', 'X-Auth-Token': stack_api.tokes}
    cloud_host_data=requests.get(request_url, headers=dc).json()
    return cloud_host_data
```

(2) app.py 部分代码如下：

```python
@app.route('/get_host_data', methods=['GET', 'POST'])
```

```python
def return_host_data():
    if request.method=="GET":
        host_data=obj.cloud_host_list()
        for i in host_data["servers"]:
            i["image"]["name"]=obj.get_image_name(i["image"]["id"])
            i["created"]=method.utc2local(str(i["created"]))
        return json.dumps(host_data)
```

(3) cloud_host_manage.html 代码如下:

```javascript
$(function () {
// 清空所有的子结点
$("#J_TbData").empty();
// 加载云主机运行数据
$.ajax({
    url: "/get_host_data",// 后台请求的数据
    dataType: "json",// 数据格式
    type: "get",// 请求方式
    async: false,// 是否异步请求
    success: function (data) {   // 如果请求成功，返回数据
        var host_list=data["servers"];// 镜像相关数据
        for(var j=0; j<host_list.length; j++) {
            var i=host_list[j];
            // 动态创建一个 tr 行标签，并且转换成 jQuery 对象
            var $trTemp=$("<tr></tr>");
            // 往行里面追加 td 单元格
            $trTemp.append("<th><input type='checkbox' name='foo' class='id' value='" + i["id"] + "'></th>");
            $trTemp.append("<td>" + i["name"] + "</td>");
            $trTemp.append("<td>" + i["image"]["name"] + "</td>");
            // 判断是否有网络
            if (JSON.stringify(i["addresses"])=='{}') {
                $trTemp.append("<td>" + "无" + "</td>");
            } else {
                for (var x in i["addresses"]) {
                    $trTemp.append("<td>" + i["addresses"][x][0]["addr"] + "</td>");
                }
            }
            $trTemp.append("<td>" + i["status"] + "</td>");
```

```
        $trTemp.append("<td>" + i["OS-EXT-SRV-ATTR:host"] + "</td>");
        $trTemp.append("<td>" + i["OS-EXT-STS:vm_state"] + "</td>");
        if (i["flavor"]) {
            $trTemp.append("<td>" + " 已启用 " + "</td>");
        } else {
            $trTemp.append("<td>" + " 已禁用 " + "</td>");
        }
        $trTemp.append("<td>" + i["created"] + "</td>");
        $trTemp.append("<td class='big'>" +
            "<div class='one'>" +
            "<a class='btn' href='javascript:void(0)' onclick=
            'fnc(this)' value='suspend'> 挂起 </a>" +
            "<div class='two'>" +
            "<a href='javascript:void(0)' onclick=
            'fnc(this)' value='start'> 开机 </a>" +
            "<a href='javascript:void(0)' onclick=
            'fnc(this)' value='stop'> 关机 </a>" +
            "<a href='javascript:void(0)' onclick=
            'fnc(this)' value='reboot'> 重启 </a>" +
            "<a href='javascript:void(0)' onclick=
            'fnc(this)' value='resume'> 恢复 </a>" +
            "<a href='javascript:void(0)' onclick=
            'rebuild(this)' value='" + i["image"]["id"] + "'> 初
            始化 </a>" +
            "<a href='javascript:void(0)' onclick=
            'remote(this)'> 远程 </a>" +
            "</div></div>" +
            "</td >");
        $trTemp.appendTo("#J_TbData");
        }
    },
});
```

2. 云主机运行相关功能

这里包含开启、关闭、恢复、重启、挂起、删除等操作。

(1) api_class.py 部分代码如下：

```
def server_manage(self, server_id, option):
host_url=stack_api.OS_AUTH_URL + ':8774/v2/servers/' + server_id + "/action"
dc={'Accept': '*/*', 'Connection': 'close', 'X-Auth-Token': stack_
```

```
api.tokes}
    # 开启
    if option=="start":
        bodys={"os-start": None}
    # 关闭
    elif option=="stop":
        bodys={"os-stop": None}
    # 恢复
    elif option=="resume":
        bodys={"resume": None}
    # 重启
    elif option=="reboot":
        bodys={
            "reboot": {
                "type": "HARD"
            }
        }
    # 挂起
    elif option=="suspend":
        bodys={"suspend": None}
    requests.post(host_url, data=json.dumps(bodys), headers=dc)
# 删除云主机
def delete_server(self, server_id):
    host_url=stack_api.OS_AUTH_URL + ':8774/v2/servers/' + server_id
    stack_api.headers['X-Auth-Token']=stack_api.tokes
    requests.delete(host_url, headers=stack_api.headers)
```

(2) app.py 部分代码如下：

```
@app.route('/servers_manage', methods=['GET', 'POST', 'DELETE'])
def servers():
    if request.method=="POST":
        server_id=json.loads(request.form.get("server_id"))
        option=request.form.get("option")
        if option=="rebuild":
            image_id=request.form.get("image_id")
            obj.server_rebuild(server_id[0], image_id)
            return "初始化成功"
        elif option=="delete":
            for i in range(len(server_id)):
```

```
                obj.delete_server(server_id[i])
            return "success"
        else:
            for i in range(len(server_id)):
                obj.server_manage(server_id[i], option)
            return "success"
else:
    return render_template("cloud_host_manage.html")
```
(3) cloud_host_manage.html 代码如下：
```
function fnc(obj) {
option=$(obj).attr("value");
lists=new Array();
server_id=$(obj).parents('tr').find('.id').val();
lists.push(server_id);
$.post("/servers_manage", {"server_id": JSON.stringify(lists),
"option": option}, function (status) {
        alert(status)
        location.href="servers_manage";
    }
);
}
```
3. 初始化，远程
(1) api_class.py 部分代码如下：
```
def server_rebuild(self, server_id, image_id):
host_url=stack_api.OS_AUTH_URL + ':8774/v2/servers/' + server_id +
"/action"
stack_api.headers['X-Auth-Token']=stack_api.tokes
bodys={
    "rebuild": {
        "imageRef": image_id
    }
}
requests.post(host_url, headers=stack_api.headers, data=json.
dumps(bodys))


def server_remote(self, server_id):
host_url=stack_api.OS_AUTH_URL + ':8774/v2/servers/' + server_id +
"/action"
```

```python
bodys={
    "os-getVNCConsole": {
        "type": "novnc"
    }
}
dc={'Accept': '*/*', 'Connection': 'close', 'X-Auth-Token': stack_api.tokes}
result=requests.post(host_url, headers=dc, data=json.dumps(bodys)).json()
# 返回远程url
return result['console']['url']
```

（2）app.py 部分代码如下：

```python
if option=="rebuild":
    image_id=request.form.get("image_id")
    obj.server_rebuild(server_id[0], image_id)
    return "初始化成功"
@app.route('/remote', methods=['GET', 'POST'])
def remote_console():
    server_id=request.form.get("server_id")
    remote_url=obj.server_remote(server_id)
    return remote_url
```